87 7

DATE DUE

ORBITING THE SUN

Planets and Satellites
of the Solar System

THE HARVARD BOOKS ON ASTRONOMY

Edited by George Field, Owen Gingerich, *and* Charles A. Whitney

ATOMS, STARS, AND NEBULAE
Lawrence H. Aller

THE MILKY WAY
Bart J. Bok and Priscilla F. Bok

GALAXIES
Harlow Shapley / Revised by Paul W. Hodge

STARS AND CLUSTERS
Cecilia Payne-Gaposchkin

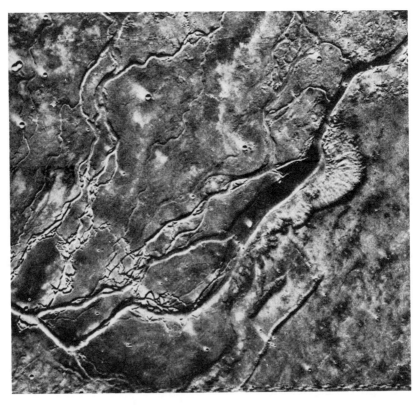

Dry rivers on Mars. Do they flow at intervals of about 120,000 years or are they purely fossil rivers? (Courtesy of the National Aeronautics and Space Administration.)

Fred L. Whipple

ORBITING THE SUN

Planets and Satellites of the Solar System

HARVARD UNIVERSITY PRESS

Cambridge, Massachusetts
London, England   1981

A new and enlarged edition of *Earth, Moon, and Planets* by Fred L. Whipple

Copyright © 1941, 1963, 1968, 1981 by the President and Fellows of Harvard College
All rights reserved
Printed in the United States of America

Library of Congress Cataloging in Publication Data

Whipple, Fred Lawrence, 1906–
  Orbiting the sun.

  (The Harvard books on astronomy)
  "A new and enlarged edition of Earth, moon, and
planets."
  Bibliography: p:
  Includes index.
  1. Planets. 2. Earth. 3. Moon. I. Title.
II. Series: Harvard books on astronomy.
QB601.W6 1981   523.2   80-19581
ISBN 0-674-64125-6

Preface

The space age mothered a renaissance of solar-system astronomy. Earthbound astronomers had become frustrated by their inability to learn much more about the planets and satellites from observations made at great distances through a turbulent atmosphere. Exciting new vistas of stellar structure and evolution, an exploding universe, pulsars, black holes, quasars, and other enticing problems whetted their taste for the green fields beyond the solar system. But with the successes of the lunar missions, the manned Apollo missions, and scientific spacecraft to and about Mars, Venus, Mercury, Jupiter, and Saturn, the balance of study between our unique backyard in the solar system and the fantastic universe beyond has been rectified.

This book is an effort to present to the general reader the major results of these space ventures, unprecedented in world history. Drawing from its predecessor, *Earth, Moon, and Planets,* it also provides the reader with a basic foundation in astronomy, through the simple presentation of scientific principles.

I regret that little or no specific recognition can be made of the multitude of scientists, engineers, administrators, and technicians who conceived, built, and operated these magic machines of space and who interpreted their coded messages. The prime contributor is the United States National Aeronautics and Space Administration, including certain of its centers, particularly the Jet Propulsion Laboratory of the California Institute of Technology. The space program of the U.S.S.R. has made several firsts. To all of them my respect and profound thanks. Also I wish to express my admiration for the people of government and for the public generally who supported these missions, the former with their status at stake and the latter with their pocketbooks.

I hope that those who read this book will share the excitement of exploration and discovery permeating these great adventures of mankind with machines. Though the hope of finding life within our solar system has dimmed, surely we will, in time, find evidence that we are not alone in the universe.

I am indebted to Fred A. Franklin for many discussions and a critical reading of the manuscript. The illustrations have greatly benefited from the artwork of Joseph F. Singarella and the photographic techniques of Charles L. Hanson, Jr. The collecting of illustrations was greatly eased by Jurrie van der Woude of the Jet Propulsion Laboratory and by Kelley Beatty of *Sky and Telescope*. For typing I wish to thank Ursulle M. Gallerani. Original materials from Michael J. S. Belton, D. B. Campbell, William K. Hartmann, Gordon H. Pettengill, James B. Pollack, Bradford A. Smith, Joseph Veverka, Richard L. Walker, and Robert A. West are gratefully acknowledged.

Cambridge, June 1980   F.L.W.

About Units of Measurement

The current trend in all countries is to use the simple and sensible metric system. Because it is universal in science, readers of astronomy should become familiar with the system. Hence I have used the metric system exclusively in this book. The major units of measurement are defined or compared in Appendix 3.

Fortunately, the units of time remain familiar to all readers. I have adopted Harold C. Urey's usage of *aeon* to designate 1,000 million years (10^9 years). Aeon is a short, suggestive word and prevents confusion with the British usage of *billion* as equal to a million million.

For distance I have tried to use *meter* whenever suitable because it can be visualized as a *yard* to 9 percent accuracy. The *kilometer* is 0.61 miles. For astronomical distances a factor of 0.6 or its reciprocal, 1.6, really doesn't make much difference conceptually. For small distances a *micron* or *micrometer* is 1/25,000 of an inch, or twice the wavelength of blue-green light. A *centimeter* is 0.4 inches.

With regard to masses (or weights) the *kilogram* is 2.20 pounds and the *metric ton* 2,205 pounds, almost identical to the *long ton* of 2,240 pounds, and only 10 percent greater than American 2,000-pound *short ton*. A *gram* is 1/28 of an ounce.

Temperatures are exclusively in degrees *Celsius* or *centigrade*, °C. Remember that 0°C is the melting point of ice, 32°F (Fahrenheit), and that the two temperature scales agree at $-40°$. Water boils at 100°C and 212°F. The ratios of scales is 5/9, the C° being larger and therefore usually smaller in numerical value. *Absolute zero,* the beginning of the *Kelvin* scale, is -273.16°C or -459.59°F.

Velocity or speed in meters per second can easily be changed to miles per hour by multiplying by 2.24 (almost exactly the factor to change kilograms to pounds).

PHOTOGRAPHS

North is usually up, unless otherwise indicated, following the spage-age custom rather than telescopic tradition.

Contents

ORBITING THE SUN

Planets and Satellites of the Solar System

1

Introducing the Planets

The five bright planets have been known to man for many thousands of years, but in antiquity they were regarded as mysterious celestial deities whose very motions seemed to reflect the caprices of superhuman beings. The old Greek and Roman legends are well known. Mars was the god of war, Venus the goddess of love, while Mercury was a sort of messenger boy. Today, the situation has changed. In the space age man-made vehicles circle the planets and land on them. Spectacular close-up pictures bring a solid reality to these bright points on the sky, revealing the planets and their moons as massive balls of iron, stone, and gas, four of which dwarf our Earth. Even our own Moon finds peers among the satellites of the giant planets. The variety among these bodies matches the variety we find on Earth, though sometimes etched on a colossal scale. Our search for living organisms beyond Earth continues, but with a weakening optimism.

These planetary spheres have real significance in our lives

and in our thinking. The mathematical calculation of orbital paths for our space vehicles is a practical engineering problem. The composition of the atmospheres, the temperature, and the nature of the planets occupy the attention of many people, not only scientists but engineers, explorers, astronauts, politicians, the military, business people, and others. Building bases in space, establishing colonies on the Moon, and mining the asteroids no longer fall under the heading of science fiction but are becoming matters of public policy. Astronomy has moved from its ivory tower into the market place and the planets have become our next-door neighbors.

Each planet and moon acquires more character and interest as the space age brings new facts. Every year some of the problems of the preceding year are solved and new ones, once beyond the expectation of solution, appear within reach. To appreciate the current discoveries and deductions in planetary astronomy, we must be familiar with the fund of knowledge already accumulated.

To proceed toward an understanding of the solar family of planets, their dependents, and the other inhabitants of this realm, an introduction is first in order. As everyone knows, the process of meeting a numerous family *en masse* may be both exhilarating and confusing. We shall proceed quickly with the introductions and then spend some time with each member of the family to develop more intimate acquaintance.

The planets are really so small compared to the vast distances between them, and their reflected sunlight is so weak in comparison with the great brilliancy of the Sun, that all of them can never show to good advantage from any one location. As a vantage spot for observations, the Earth is actually quite satisfactory, except for the thick atmosphere above us. Since we must surmount this obstacle, we might as well go out farther from the Sun to about the distance of Jupiter. From there the inner planets of the solar system would appear somewhat like the sketch in Fig. 1. First we notice that the Sun is almost exactly in the center of the planets' orbits. The reason is very simple: the Sun possesses 99.866 percent of the entire mass of the system, so that by gravitational attraction it completely dominates the motions of the planets.

We notice next that the orbits lie almost in a plane, very close

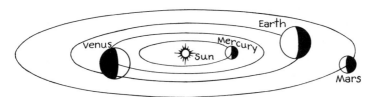

Fig. 1. Orbits of the inner planets about the Sun—a projection. The relative sizes of the planets are indicated, but on the scale used, the Sun's diameter would be 1 meter.

to the *ecliptic,* the plane of the Earth's orbit about the Sun. The orbit of Mercury, the smallest planet, is tipped from the common plane with an inclination of 7°, but the other inner planets keep within about 3° of the plane. This favoritism of the planets toward a common plane of motion cannot be due to chance. Although no rigorous proof has been given, it is possible that Jupiter is responsible, because this planet is 318 times as massive as the Earth and possesses 0.7 of the combined mass of all the planets. Jupiter is certainly the master planet and by gravitational attraction may have regulated the orbits of the others. There is the more likely possibility, of course, that the planets were all formed in a plane—a matter for later investigation.

For measuring distances in the solar system the most convenient unit is the *astronomical unit* (AU), which equals the mean of the greatest and least distances from the Earth to the Sun, technically called the Earth's *mean distance.* This basic yardstick is some 149,597,900 kilometers long, by astronomical and radar measures. The distance is still uncertain by about a hundred kilometers. To get an idea of how large an astronomical unit is, we could calculate that the time required for an airplane moving with the velocity of sound, 1200 kilometers per hour, to travel from the Earth to the Sun is approximately 14 years (necessarily a one-way trip), while a rocket moving at 10 kilometers per second would arrive in 6 months. The astronomical unit is actually much too small for conveniently listing the distance between the stars; a much larger unit, the distance light travels in a year, is often used for that purpose. This unit, as well as numerical constants for the planets, is given in Appendix 3.

Mercury has a mean distance from the Sun of only 0.39 AU, Venus 0.72, the Earth 1.00, Mars 1.52, and Jupiter 5.20. (A convenient scheme for remembering the distances of planets, Bode's law, is given in Appendix 1.) This sequence of increasing distances is rather uniform except for the large gap between Mars and Jupiter. In this gap we find thousands of small planets called *asteroids* that fill the space where a planet might well move (see Fig. 2 and note Fig. 7, later). These asteroids range from mountain size, less than a kilometer in diameter, up to Ceres, which is some 1003 kilometers across—comparable to a large island. Pallas comes second with a diameter of 610 and Vesta

Fig. 2. The famous asteroid Eros leaves a trail on this time exposure as it moves through a field of stars. (Photograph by the Yerkes Observatory.)

third, 540 kilometers. There are certainly no large asteroids that have not been discovered but there are many smaller ones —millions—that can be photographed with the larger telescopes. These fly-weight planets, although contributing a negligible part to the mass of the system (perhaps $\frac{1}{500}$ of the Earth's mass), provide astronomers with a great amount of work in observation and calculation. They are fine test specimens for theories of various kinds and are a major key to the origin of the solar system. Probably they were formed much in their present positions although collisions have certainly scarred and battered them, if not indeed decimated them.

The planets themselves show much of the character of the ancient gods for whom they were named. Mercury is indeed swift and small, characteristic of a messenger. It requires only 88 days for a complete revolution about the Sun, less than one-fourth the length of our year. Its diameter is only 0.4 that of the Earth. Even this small diameter, 4878 kilometers, is enough greater than the diameter of Ceres to establish Mercury definitely as a planet rather than as a large asteroid. On the other hand, some of the moons of the giant planets match Mercury in size. Radar first showed that Mercury rotates in about 59 days. The U.S. Mariner 10 space probe shows that Mercury is gravitationally locked into exactly two-thirds of its period of revolution about the Sun, 58.65 days. The planet is, unfortunately, so small and always remains so close to the Sun, as observed from the Earth, that surface markings are difficult to discern. The magnificent Mariner 10 television pictures of Mercury (Fig. 3) show a striking similarity to those of the Moon. These airless worlds bear the scars, perhaps even birthmarks, of countless collisions.

Venus is certainly the "sister" planet of the Earth. The diameter is almost identical (95 percent), the period of revolution about the Sun somewhat shorter (225 days), and the mass about 0.8 that of the Earth. Venus too is cloaked with a large atmosphere; it is this opaque atmosphere that hides the surface features so completely that we could not even determine the direction of rotation before the days of interplanetary radar. The period is 243 days, and *retrograde,* opposite to the direction of revolution about the Sun. Radio first, then the USSR and later the U.S. space probes, show that the surface temperature is some 457°C, too hot even for sulfur, which cools to droplets and

Fig. 3. Mosaic of southern hemisphere of Mercury, imaged by Mariner 10 in 1974. (Courtesy of the National Aeronautics and Space Administration.)

dust along with sulfuric acid droplets in the clouds at high altitudes. The weight of ninety Earth atmospheres, mostly carbon dioxide, blankets the planet.

Earth-based observations of both of these inner planets are difficult because we can see only part of the sunlit face at any time. When nearest to the Earth, Venus shows a thin crescent, as is the case for the Moon when new or old, because the bodies

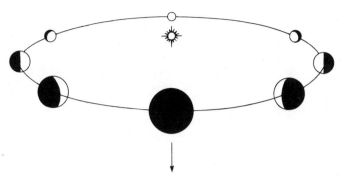

Fig. 4. The phases of Venus relative to the Earth. See Fig. 5 for corresponding photographs.

are almost on a line between the Earth and the Sun. Figure 4 shows the various positions of Venus when the photographs of Fig. 5 were made. The various planetary configurations are named and defined in Appendix 2.

Mars may best be described as a pygmy Earth (one-half its diameter) with a thin atmosphere, distinct surface features, but no oceans. It moves more slowly about the Sun, in 687 days. Mars, however, boasts two moons, while Mercury and Venus have none. These two satellites are little more than medals for Mars, the god of war, since the larger, Phobos, shaped like a badly formed potato, is only 22 kilometers across its average axis. Deimos has about one half the diameter of Phobos. Close-up pictures made by the Mars Viking mission appear in Fig. 6. Small asteroids would probably look much the same.

The existence of these Lilliputian satellites, strangely enough, was heralded by Jonathan Swift (1667–1745) in his *Gulliver's Travels* some 150 years before their discovery in 1877. According to Gulliver's account, the astronomers of the cloud island Laputa possessed small but most excellent telescopes and had "discovered two lesser stars, or satellites, which revolve about Mars; whereof the innermost is distant from the center of the primary planet exactly three of his diameters, and the outermost, five; the former revolves in the space of ten hours, and the latter in twenty-one and a half." These periods of revolution are remarkably close to the truth, for Phobos revolves about Mars in 7 hours 39 minutes, while Deimos requires 30 hours 18 minutes. The mythical distances from the center of Mars are,

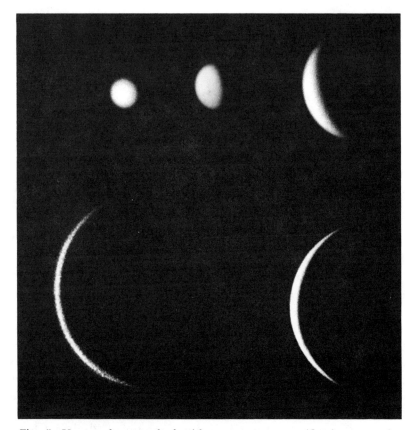

Fig. 5. Venus photographed with a constant magnification at various phases. (Photographs by E. C. Slipher, Lowell Observatory.)

however, too great; Phobos is distant only 1.4 diameters of the planet, and Deimos 3.5 diameters. It would be enlightening to learn more of the Laputian discoveries, but Gulliver mentions only that the Laputians had "observed ninety-three different comets, and settled their periods with great exactness."

Until recently it was believed that the rapid motion of Phobos made this satellite unique in the solar system. Its period of revolution is less than the Martian day, 24 hours 37 minutes. As seen from the surface of Mars, Phobos rises in the west and sets in the east! Alas, Phobos is no longer unique. The unnamed fourteenth satellite of Jupiter, discovered by NASA's Voyager in 1979, revolves about Jupiter in only 5 hours, or one-half the period of the planet.

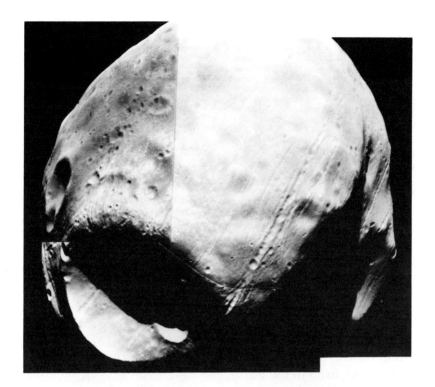

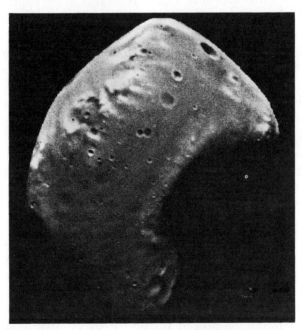

Fig. 6. The satellites of Mars. (*Top*) Phobos, the inner; (*bottom*) Deimos, the outer. Imaging by Viking. (Courtesy of the National Aeronautics and Space Administration.)

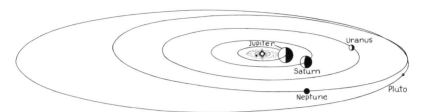

Fig. 7. Orbits of the outer planets about the Sun—a projection. Pluto passes within Neptune's orbit, but does not intersect it because of the inclination. Note the asteroids and the relatively small size of the orbit of Mars.

Before going on to the outermost planets, we note that the three planets Mercury, Venus, and Mars are really very much like the Earth, of somewhat the same size, and all fairly dense, as though they were made of stone and iron. They, along with the planet Pluto, are referred to as the *terrestrial* planets because of their similarity to the Earth. Jupiter, Saturn, Uranus, and Neptune, on the other hand, are of an entirely different species, giants compared to the Earth, and only about as dense as water. Their orbits are shown in Fig. 7 as seen from beyond the orbit of Pluto. On this chart the orbits of Fig. 1 are all crowded into a small region about the Sun. The space missions tell us so much about the planets that a new field of study has blossomed forth, *comparative planetology,* really the main subject of this book (see Chapter 14).

Jupiter stands out as the greatest of the planets. The intricate Voyager missions of NASA to the outer planets have sent back a fantastic record of this majestic system, a solar system in miniature but still colossal compared with our Earth and Moon. In the center is Jupiter, with 11 times the diameter of the Earth and 330 times its mass. Its equator bulges with the rapid rotation period of less than 10 hours. The complexities of its outer circulation pattern are beyond description but can be vaguely realized from the television images of the cloud pattern and motions near the top of the bottomless (?) atmosphere (Fig. 8, Plates I and II). Hydrogen is the major gas, with some 18 percent by mass of helium and contamination by ammonia (NH_3) and methane (CH_4 marsh gas). Powerful lightning bolts, suitable for Jupiter the god of sky and weather, strike among the clouds. A great meteor has been seen in its atmosphere.

Jupiter's gravity controls four great satellites comparable to

Fig. 8. Jupiter as imaged by Voyager 1, February 5, 1979. Io, the innermost Galilean satellite, appears against the disk. (Courtesy of the National Aeronautics and Space Administration.)

Fig. 9. Neptune and Triton. (Photograph by the Lick Observatory.)

our Moon, first discovered by Galileo with the newly invented telescope in 1610. The extraordinary Voyager pictures of these moons show them each to be unique and each a challenge for the scientist to explain (Plates III and IV). The innermost of the great satellites, Io, shows active volcanos, the first ever to be detected outside the Earth. Jupiter also controls at least ten small satellites, two—XIV and Amalthea V—moving inside Io's orbit, with XIV only 30,000 kilometers from the surface of the planet. Not surprisingly, Jupiter has the strongest magnetic field in the solar system, which enables the planet to emit strong radio noise. This radio information in turn tells us, perhaps surprisingly, that much of its interior rotates as a solid, faster than the high clouds we see.

This preview of the planet Jupiter is generally descriptive of the other giant planets, Saturn, Uranus, and Neptune, except that they are less massive, farther from the Sun, and, to date, less well studied. Neptune, the outermost at 30 AU from the Sun, is a frigid world by our standards because it receives only $\frac{1}{900}$ as much heat and light from the Sun as we receive. Solid carbon dioxide ("dry ice") near its melting point is hot compared to the temperature in the clouds of Neptune, below $-200°C$. Nitrogen gas would be frozen, likewise oxygen.

Although the giant planets are cold and uninhabitable, their great masses and wide separation in space allow them all to control astonishingly large families of satellites. Jupiter is first, with fourteen moons, as we have seen. Saturn is second with nine (or ten and probably more), while Uranus has five and Neptune only two (Figs. 9 and 10). Saturn's Titan and Neptune's Triton are comparable in size to our Moon, while the others range in diameter from that of small asteroids to about half that of the Moon. The great planets are more massive when compared with their largest satellites than is the Sun when compared to Jupiter or Saturn.

The similarity with the whole solar system is even more striking in the system of Saturn because this planet not only controls ten satellites, more than the number of known planets about the Sun, but also possesses a family of miniature asteroids, which comprise the great rings (Fig. 11). These rings are so close to Saturn itself that in the poor early telescopes they looked like ears or appendages. Galileo sometimes drew Saturn as consist-

Fig. 10. Saturn and its satellites photographed December 8, 1966 when the rings were on edge. From top left to lower right, the images are of Rhea, probably Janus (on left), Tethys (almost lost in Saturn's glare), Enceladus, Dione, and Titan. (Official U.S. Navy photograph, by Richard L. Walker, Jr., with the 61-inch reflector at Flagstaff, Arizona.)

ing of three pieces—a central body with symmetric side sections (Fig. 12). We know now that the rings are made of small particles covered with ice and revolve about Saturn in a plane that is relatively thinner with respect to its width than a sheet of paper. When seen at different angles they present different appearances, from almost invisibility, if seen edge on, to wide rings.

A long-standing question has recently been resolved: why should Saturn be the only giant planet with rings? In fact, Saturn is *not* unique. Both Uranus and Jupiter have rings, even though they are much fainter and difficult to detect. No one will

Fig. 11. Saturn, the ringed planet. (Photograph by the Mount Wilson and Palomar Observatories.)

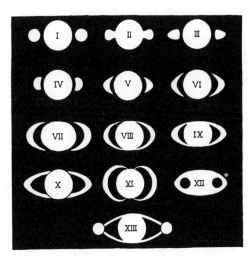

Fig. 12. Early drawings of Saturn. (From T. E. R. Phillips and W. H. Steavenson, eds., *Splendour of the Heavens*, 1923; courtesy of Hutchinson and Co., London.)

Fig. 13. Head of Comet Whipple-Bernasconi-Kulin, February 28, 1943. (Photograph by H. Giclas, Lowell Observatory.)

be much surprised if such rings are discovered about Neptune. In fact, new rings and new faint satellites are continuously being sought and found. No one knows how many tiny companions may circle the giant planets.

On beyond Neptune lies the more recently discovered Pluto. Little is known about Pluto as a planet. Although it is about Moon size, it probably is not a rocky type of body, perhaps just a mass of ice and dirt, like a comet, entirely inactive at some 40 astronomical units from the Sun. It rotates in 6 days 9 hours and shows solid methane on its surface. Surprisingly, Pluto appears to have a satellite, detected in 1978 by J. W. Christy at the U. S. Naval Observatory as an occasional slight distortion of Pluto's nearly starlike image. Appropriately named Charon for the boatman who operates the underworld ferry across the river Styx, the satellite may be nearly half the size of the planet, revolving with the period of Pluto's rotation. Hence Pluto may

Fig. 14. Two photographs of Halley's comet. The telescope was made to follow the comet's motion during a time exposure and hence the star images are
trailed. (Photographs by the Lick Observatory.)

compensate for its inferior size by the distinction of being a
double planet.

To complete this quick introduction to the solar family, some
mention must be made of the remarkable comets. These
strange wanderers have excited more superstitious fear in the
human mind than any other class of celestial bodies. Most

comets move in exceedingly elongated orbits, approaching the Sun for only a short time each revolution. When distant from the Sun they become too faint for observation, but they brighten enormously at perihelion, when closest to the Sun. At this time they become so active that they waste an appreciable part of their substance into space, and produce a great coma of gases and small dust particles about their nuclei (see Fig. 13). Sunlight and gases blown out from the sun force the cometary gases and dust back from the head in a great tail, sometimes a multiple tail, and always complicated in structure. The structure and brilliancy of the tails are obvious from inspection of Fig. 14. Astronomers now accept my theory that comets are fundamentally balls of ice and dirt and that they become active only when so close to the sun that the ice vaporizes. Although most comets are small, only a very few kilometers in diameter, they are very numerous and played a critical role in forming the outer planets, especially Uranus and Neptune. Perhaps they contributed the light volatiles such as carbon and water that make life possible on Earth.

In later chapters we shall become much better acquainted with the members of the solar family. Each has a character more appreciated on closer contact. Also there are many provocative problems of structure, origin, family relationships, and even some family skeletons. In the next chapter we shall look into the important problem of family unity, the binding force that keeps each member in its place.

2

How the System Holds Together

A mighty and all-pervading force enables the Sun to hold the planets in their orbits of revolution and empowers the planets to retain their satellites. The discovery of the law of universal gravitation, which describes this force, stands as a monumental feat of the human mind. Only a genius like Sir Isaac Newton (1642–1727) could have started with the observational material and theories of his day, developed a new form of mathematics to solve the dynamical problems, and finally welded the observations and mathematical theory to form a simple yet universal law. A better understanding of his achievement can be obtained by glancing backward at the scientific foundation from which he started.

During the two centuries preceding Newton's activities, a few European scientists had been amassing evidence and arguments to disprove the concept that the universe is centered on the Earth, benevolently lighted by the Sun with the Moon, planets, and stars as cheerful decoration. Nicholas Copernicus (1473–

1543) deserves credit for bringing into disrepute the idea of a fixed Earth, an idea long cherished by the followers of Aristotle (384–322 B.C.). Actually many ancient Greeks favored the philosophic concept of a moving earth, but Aristotle's opinion carried great weight. An early impression of planetary motions as seen from a fixed Earth is shown in Fig. 15. This system of Ptolemy's (2nd century A.D.) represented the known facts extremely well, and simply.

Once the concept of a moving Earth was recognized as likely, although not well proved, the subsequent task of finding out how the Earth moves and why it moves was still difficult. The stars, actually distant suns, are too far away to indicate by their yearly oscillation the 300-million-kilometer swing of the Earth about the Sun, even by measures made long after the invention of the telescope. One can very well sympathize with the critics of the new theory who stoutly maintained that solid earth was solid earth, obviously fixed in space. "If it were moving, as those young upstarts would have one believe, why do not the stars swing back and forth across the sky during the year?" The argument was absolutely sound, and was disproved only during the nineteenth century by means of the most accurate observing techniques. The nearest star, Proxima Centauri, is 270,000 astronomical units distant. From it the Earth's orbit would appear to have a radius smaller than the diameter of a human hair as seen 15 meters away from the eye. Thus the yearly oscillation of

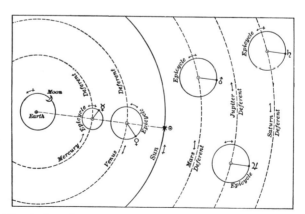

Fig. 15. The Ptolemaic system. According to the ancient Greco-Egyptian astronomer Ptolemy, the planets moved in small circles about fictitious planets that moved in large circles about the fixed Earth. (From C. A. Young, *A Text-Book of General Astronomy*, 1888; courtesy of Ginn and Co.)

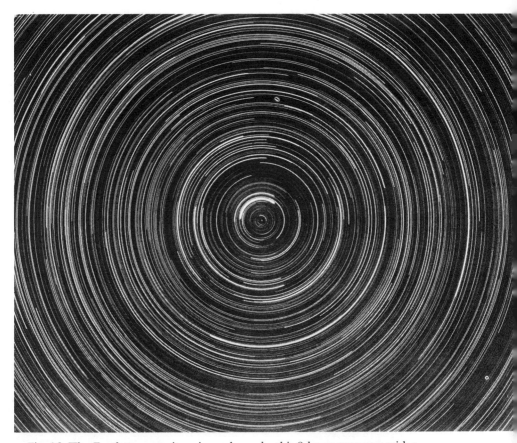

Fig. 16. The Earth turns on its axis, as shown by this 8-hour exposure with a fixed camera pointing at the North Pole. The heavy trail near the center was made by Polaris. (Photograph by Fred Chappell, Lick Observatory.)

Proxima Centauri is an angle less than the apparent motion of the hair if moved through twice its diameter. For all other stars the annual motion is even smaller.

While the argument was raging about the motion of the Earth, the difficulty of predicting future positions for the Sun and planets, within the increasing degree of accuracy to which they could be observed, was becoming more and more serious. The invention of the clock accentuated the need for better predictions and more effective instruments to measure directions on the sky. It was necessary to know accurately how the planets actually move through space. The daily rotation of the Earth (see Fig. 16) and its yearly revolution, as we know today, compli-

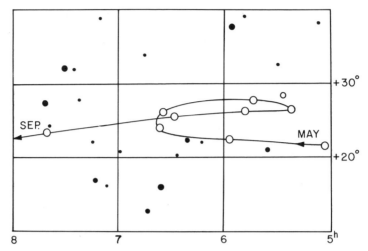

Fig. 17. Mars followed this path against the background of stars for 8 months. The circles represent positions at month intervals.

cate the problem enormously because the observations must all be made from the Earth, a body itself in motion. In addition, light rays must pass through the atmosphere, which can bend them as much as half a degree when near the horizon.

The effects of rotation and atmosphere can largely be removed if one establishes the relative positions of the stars in a fixed system covering the entire sky and then measures the positions of the planets with respect to the stars. The apparent motion of Mars during one *opposition* (see Appendix 2 for definition of planetary configurations) from the Sun is shown in Fig. 17. This peculiar curve on the star background little resembles the smooth curve of the actual space motion already shown in Fig. 1.

In the sixteenth century the great Danish astronomer, Tycho Brahe (1546–1601), set about doing what he could to improve the knowledge of planetary motions. His principle of action is one that should always be remembered by every scientist, because it embodies the very essence of good science. Tycho Brahe made the best instruments he could, made the best observations possible with them, and then carefully studied his instruments to determine the size of the errors that should be expected. His long series of observations of Mars were minutely analyzed by Johannes Kepler of Württemberg (1571–1630), who experimented with every kind of motion that he could devise for the planet. Some types of eccentric motions of Mars about the Sun would fit the observations *almost* as well as they should, but Kepler was obsessed by scientific ideals. His perseverance led him finally to discover three very simple laws describing the motion of a planet about the Sun. A simple law, if it fits the observations well, always pleases and encourages a scientist. Kepler was certain that he had learned the truth about planetary motions, and time has corroborated his opinion. His laws also describe the motions of space vehicles when they move unpowered between the planets as well as the motions of satellites about planets.

Kepler's first law states that *the orbit of a planet is an ellipse with the Sun at one focus.* Now an ellipse is one of the simplest closed curves on a plane, one that has always delighted mathematicians because of the many simple theorems to which it is susceptible. To obtain an ellipse is nearly as easy as to draw a circle. Simply take a cone (right circular, if a mathematician is nearby) and slice it with a plane. The curve where the two surfaces intersect is an ellipse (Fig. 18*b*). You may, of course, be ingenious and

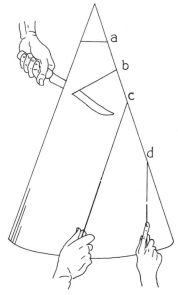

Fig. 18. A cone is sectioned by a plane to produce (*a*) a circle, (*b*) an ellipse, (*c*) parabola, or (*d*) a hyperbola. These plane curves are called conic sections.

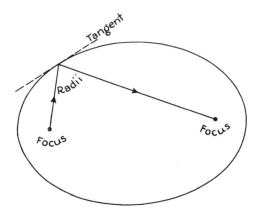

Fig. 19. An ellipse. Radii from the two foci make equal angles with a tangent. The eccentricity of this ellipse is 0.72.

pass the plane through the vertex to obtain only a point, or perpendicular to the axis to obtain a circle *(a),* or parallel to a side. In the last mentioned case the ellipse never closes, becoming a parabola *(c),* or even a hyperbola *(d),* if the plane is more nearly vertical. These possibilities are no problem to the mathematician, who calls all the possible curves *conic sections* and then proceeds to derive even more general theorems about them.

There are a number of theorems about the focus of an ellipse. For example, if we draw a line from one focus and reflect the line from the ellipse at an equal angle to the tangent, the line will always pass through the second focus, as in Fig. 19. The situation is even simpler for a parabola where rays from the focus are reflected as a parallel bundle. This is the principle of a searchlight or an automobile headlight. In reverse, it is the principle of the reflecting telescope where parallel light rays from a distant star are brought to a point focus by reflection from the surface of a parabolic mirror (Fig. 20).

Another noteworthy property of an ellipse is that from any point of the ellipse the sum of the distances to the two foci is constant. This property suggests a very easy method of drawing an ellipse. Stick two strong pins into a sheet of paper at the points where the foci are to be. Then place a closed loop of string about the two pins, stretch the loop taut with the point of a pencil, and draw the ellipse by swinging the pencil around the

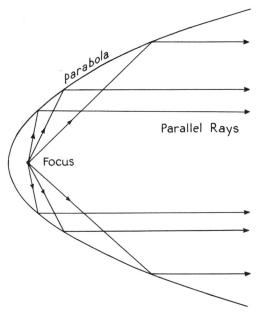

Fig. 20. A parabola. All radii from the focus are reflected into parallelism by a parabola.

pins inside the taut loop (Fig. 21). If the two pins are together one draws a circle, the simplest ellipse.

According to Kepler's first law, the Sun is always at one focus of the ellipse, the other focus being empty—a completely neglected mathematical point. Various possible orbits are drawn

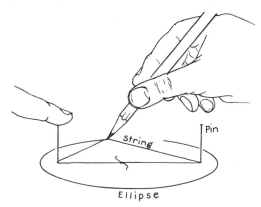

Fig. 21. Drawing an ellipse by means of two pins and a loop of string. This method works well except for the knot. It is better to tie the string at one pin.

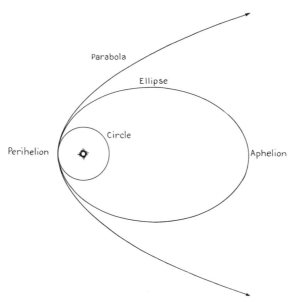

Fig. 22. Various orbits about the Sun. Comets follow orbits that are very elongated ellipses, nearly parabolic close to the Sun. Planetary orbits are ellipses but nearly circular.

in Fig. 22. The point nearest to the Sun is *perihelion* and the point farthest away is *aphelion*. The *mean distance* is half the sum of the perihelion and aphelion distances, or the semimajor axis of the ellipse. The shape of the orbit is measured by the *eccentricity,* which is the difference of the aphelion and perihelion distances divided by their sum. The eccentricity is zero for a circle, 1.0 for a parabola, and about 0.5 for a man's hatband.

The Earth's orbit is almost a circle, having an eccentricity of only $\frac{1}{20}$. To the eye such an ellipse resembles a circle fairly well drawn, but the focus is clearly not in the center. Mercury and Pluto are the only planets with orbits that deviate much from circles, their eccentricities being 0.21 and 0.25, respectively. The distance of Pluto from the Sun thus varies from 30 astronomical units at perihelion, just less than Neptune's mean distance, up to 50 at aphelion. One checks this calculation by noting that the mean of 30 and 50 is 40, the mean distance of Pluto in astronomical units, and noting that $(50 - 30)$ divided by $(50 + 30)$ is $\frac{20}{80}$ or 0.25, the eccentricity.

Kepler's second law of planetary motions is simpler than the

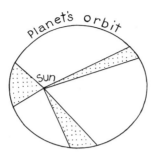

Fig. 23. Equal areas are swept out in equal intervals of time. A planet moving in this orbit would pass through each of the shaded areas in the same time. The eccentricity of the ellipse is 0.5.

first. It states that *a line joining a planet to the Sun sweeps out equal areas in equal intervals of time.* Accordingly, when a planet is near the Sun at perihelion it must move at a greater speed than when it is farther away, say at aphelion (see Fig. 23). For Pluto the speed is 6.1 kilometers per second at perihelion and 3.7 kilometers per second at aphelion. The ratio is $\frac{5}{3}$, as we might have guessed from the ratio of the two distances. At perihelion (about January 1) the Earth has increased its speed by 1.0 kilometers per second from its aphelion speed of 29.3 kilometers per second.

Kepler's third law, the harmonic law, states that *the squares of the periods of revolution of the planets about the Sun are in the same ratio as the cubes of their mean distances.* This law provides an easy method for calculating the period if one knows the mean distance of a body revolving about the Sun. Express the mean distance in astronomical units. Cube this distance. The square root of the cube is the period of revolution in years. For the Earth the formula checks from the definition of the astronomical unit and the year: the square root of 1^3 is 1; the period of the Earth is 1 year. For Neptune the mean distance is 30 AU, and $30^3 = 27,000$. The square root of 27,000 is 164, Neptune's period in years. More accurately, the period is 164.8 years, obtained by using a more precise value of the mean distance.

By means of Kepler's three laws, the elliptic law, the law of areas, and the harmonic law, a planet's motion can be predicted far into the future. Only three corrections are applied today. The masses of the planets change the laws slightly; one planet will disturb the motion of another; and a slight correction must be made in the case of Mercury's orbit because of an effect predicted by Einstein's theory of relativity.

Newton was fully conversant with Kepler's laws for describing the motions of the planets and also with Galileo Galilei's (1564–1642) revolutionary idea that all bodies fall at the same rate regardless of mass or weight. Galileo is said to have demonstrated this idea by dropping large and small weights from a tower, which may have been the Leaning Tower of Pisa. Galileo held another revolutionary idea, that in space bodies would continue to move indefinitely unless stopped by some force. But Galileo's well-known difficulty with the Church arose largely from his teaching Copernicus' theory that the earth really moves.

Newton extended Galileo's ideas about the motion of material bodies in empty space and crystallized them into three simple laws. These principles of motion are so familiar to us today that they are listed only for the sake of completeness. The first states that *a body remains at rest or maintains a uniform motion in a straight line unless acted upon by a force;* the second states that *the rate of change of motion is proportional to the force acting* (really a definition of *force*); and the third states that *action and reaction are equal but opposite in direction.* Obvious applications exist everywhere in our modern world of machines. Lack of measurement of frictional forces, both with the air and between moving parts in machinery, is the one difficulty that prevented the laws from being discovered much sooner.

With all of these principles in mind, Newton began to ponder the problems of the motions of the Moon and the planets. Since by gravitation the Earth attracts an apple, or a cannonball, or a feather, each with a force proportional to the mass, why should it not attract the Moon? By all rights the Moon should move in a straight line unless it is acted upon by a force—but the Moon actually moves in a curved path about the Earth. Therefore, it must be continually falling toward the Earth, the rate of fall being measured by the deviation from motion in a straight line (Fig. 24). Thus, the attraction of the Earth must produce a force on the Moon of exactly the right magnitude to cause the Moon to fall as it does.

"But how does the force of gravity decrease with distance from the earth?" asked Newton. To answer this he first calculated the law of centripetal force, the central force exerted

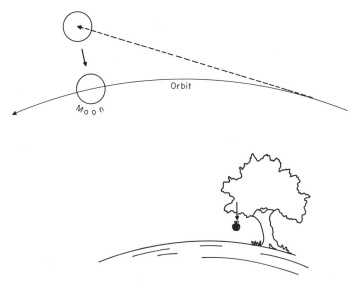

Fig. 24. Falling Moon and falling apple. Newton concluded that the Earth's gravity causes the Moon to fall toward the Earth from a straight line and the apple to fall from a tree according to the same law.

when a ball is whirled at the end of a string fastened at a point. To produce this centripetal force he found that gravity should fall off as the inverse square of the distance to the attracting center. Then from Kepler's laws he independently deduced that the planets must also be attracted toward the Sun by an inverse-square law of force.

Newton was then ready to check his theory by the Moon's motion. Here he ran into difficulties. He first used an erroneous value of the Earth's dimensions and also found some trouble in proving that a spherical Earth should act gravitationally as though its entire mass were concentrated at its center. Finally, however, the check worked. Putting all the evidence together he came to the conclusion that *every particle of matter in the universe attracts every other particle with a force that varies directly as the product of their masses and inversely as the square of the distance between them.* This law of universal gravitation accounts for all the complicated motions in the solar system to the high degree of accuracy possible in astronomical measurement (about one part in a million). The only error is a forward motion of the perihelion of Mercury, by about 43″ in a century, which is explained by a slight correction to Newton's law as predicted by Einstein's theory of relativity. An angle of 43″ would be subtended by the iris of the eye at a distance of about 50 meters.

Thus the solar system holds together by the attraction of the Sun upon the planets, and the satellite systems by the attraction of the planets upon their satellites. The problem would be very simply and completely solved by Kepler's laws if it were not for the unhappy circumstance that all of the planets attract each other, as well as their satellites and the Sun. This universal attraction complicates the problem so much that there is no exact mathematical solution. The only saving grace lies in the fact that the planets are much less massive than the Sun so that the forces of interattraction, proportional to the masses, are much smaller than the attraction of the Sun. The satellites, likewise, are much less massive than their planets. Consequently, Kepler's laws can be used to obtain an approximate solution for the motions, small corrections being made on the basis of the interattractions. These corrections are known as *perturbations,* because the other planets *perturb* the motions of the one under investigation.

The most difficult classical problem of perturbations occurs in the Earth-Moon system where the Sun perturbs the motion of the Moon about the Earth and where our observations are so excellent because the Moon is so close. Strictly speaking, the Earth perturbs the motion of the Moon about the Sun because the Sun actually attracts the Moon with a force nearly twice as great as the attraction of the Earth on the Moon. Nevertheless, there is no danger that the Sun can steal the Moon away from the Earth and leave us without inspiration on warm summer nights, at least not for some hundreds of millions of years. The system is so compact, the Earth and Moon moving so nearly together, that the Sun's attraction serves only to keep both bodies moving about it in an average path that is elliptic. The chief results are, first, that the Moon's orbit is never convex *toward* the Sun (Fig. 25), and second, that astronomers have much more work to do in predicting the Moon's motion. One single equation for the motion of the Moon covers some 250 large-size pages and represents the major effort of a lifetime—a lifetime, that is, before the advent of the modern electronic computer. Now the computer can derive the necessary series expansions as well as provide the numerical answers.

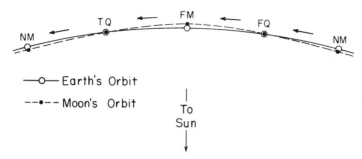

Fig. 25. The Moon's orbit about the Sun. The deviations from a perfect ellipse are greatly exaggerated in the diagram; even so, the Moon's orbit is always concave toward the Sun, as can be seen by tipping the page and sighting along the broken line.

The fact that the Earth is not exactly a sphere and does not attract precisely from its center adds a further slight complication to the motion of the Moon. The motions of artificial Earth satellites, however, are much disturbed by this effect, and, surprisingly enough, by the pressure of sunlight and even radiation from the Earth. Furthermore, the resistance of the Earth's high atmosphere tends to haul them down. Their periods of revolution, too, are so short—a minimum of only 88 minutes— that all these complicated effects must be calculated rapidly, sometimes in minutes. Thus electronic computers, millions of times faster than older methods of calculating, are absolutely essential to the success of our modern satellite and space-vehicle programs, and a godsend to almost all aspects of science today.

Among the planets, Jupiter is definitely the solar system's bad boy in disturbing the motions of all the planets and asteroids. With a mass equal to 0.001 that of the Sun, a lion's share of the mass of the entire planct family, Jupiter produces by far the largest perturbations, particularly among the asteroids which are nearest to it in space (Fig. 7). If the orbit of an asteroid is calculated without allowing for Jupiter's attraction, the errors of prediction may amount to several degrees within a few years. Asteroids sometimes get "lost" in this fashion, until they are independently rediscovered and identified by their orbital path and brightness.

Jupiter also affects the motions of the asteroids systematically as shown by the distribution of periods, or semimajor axes (Fig. 26). Conspicuous gaps appear when the periods are certain

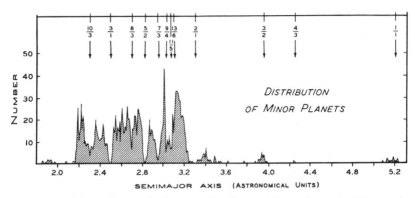

Fig. 26. Kirkwood's gaps in the periods of asteroids are shown in this compilation. The fractions are the ratios of the period of Jupiter to that of the asteroids. (Courtesy of *Sky and Telescope.*)

fractions of Jupiter's period such as $\frac{1}{3}$, $\frac{2}{5}$, $\frac{3}{7}$, and $\frac{1}{2}$. But a clump appears at $\frac{2}{3}$. The detailed explanation is actually very complicated, involving interactions among the asteroids. Again in Saturn's rings, which we have mentioned as similar to the asteroid system, exactly the same effect occurs. The fact that there are *rings* rather than a single ring results from the perturbations of the minute moonlets in the rings produced by three inner satellites, Mimas, Enceladus, and Tethys. These satellites force the small particles in the rings to avoid certain speeds in moving about Saturn, with the result that dark spaces show at the critical distances. Cassini's division (G. D. Cassini, 1625–1712), the conspicuous dark space between the outer and middle rings (Fig. 27), covers a region in which the tiny moonlets would move with periods of one-half that of Mimas, one-third that of Enceladus, or one-quarter that of Tethys, slightly more than 11 hours. Other expected "dark rings" can also be observed.

In this chapter we have considered Newton's law of gravitation as a *force* between masses. Albert Einstein's (1879–1955) general relativity improves Newton's law by a small correction but, far more importantly, entirely alters the basic concept. No longer can we accept *action at a distance* or a *force* as philosophically valid. In Einstein's concept any mass distorts, curves, or warps the space–time continuum about it. The moving masses then follow paths that are exceedingly close to those predicted

Fig. 27. Saturn's rings as drawn by B. Lyot. (Courtesy of A. Dollfus.)

by Newton's law. In the solar system these deviations are rarely measurable so that the discussion of this chapter, based on the Newtonian concept, is a practical working model, an approximation to physical reality. The philosophical significance of relativity is, of course, an important subject for study by the ambitious reader.

The fact that there exists such a large ratio of mass between the Sun and the planets, and between the planets and their satellites, cannot be just a chance circumstance. The solar system would be drastically different were the planets comparable in mass with the Sun. No one of the planets could stay near the center of gravity, as the Sun does now in the solar system, but all would be moving in complex curves of almost unpredictable form. Although mathematics cannot provide a useful solution even for the case of three bodies of nearly equal mass, it demonstrates that in our hypothetical case there would be disastrous results. Some planets would be eliminated by collisions and others tossed out of the system until, most probably, the system would eventually consist of two of the largest bodies moving about each other at a moderate distance, with smaller companions or satellite systems. Other bodies might remain in the system at very great distances from the largest ones. In stable systems each subsystem, for example Jupiter and its satellites, must be adequately separated from the other subsystems so that each is only mildly disturbed by the others. Our experience with double and multiple stars indicates that they tend to occur in pairs separated by relatively large distances from other pairs in the system. There is no danger that the solar system will lose any planets or that any major collisions will occur for hundreds of millions of years, a comforting thought.

3

The Discoveries of Neptune and Pluto

The remarkable story of the discoveries of Neptune and Pluto begins actually with the discovery of Uranus, because without observations of Uranus the two later discoveries would have been delayed for many years. Indeed, the detection of Uranus marks the beginning of a new epoch in the history of astronomy, for Uranus was the first planet to be "discovered." Mercury, Venus, Mars, Jupiter, and Saturn have always been visible to the eye of anyone who looked skyward (unless the eyes of our prehistoric ancestors were much inferior to our own).

Sir William Herschel (1738–1822; Fig. 28), perhaps the most assiduous observer of all time, first detected the small disk (3.6 seconds of arc) of Uranus in 1781. His account of the discovery shows clearly that he was not immediately certain of the true character of the new object. From the *Philosophical Transactions* of 1781 we read, "On Tuesday the 13th of March, between ten and eleven in the evening, while I was examining the small stars in the neighborhood of H Geminorum, I perceived one that ap-

Fig. 28. Sir William Herschel.

peared visibly larger than the rest; being struck with its uncommon magnitude, I compared it to H Geminorum and the small star in the quartile between Auriga and Gemini, and finding it to be much larger than either of them, suspected it to be a comet."

Herschel's announcement of the new object as a comet was natural and conservative, whatever he may have suspected concerning its true nature. Several months of observation and calculation were required to demonstrate that no cometary motion would satisfy the observations and that the "comet" could be nothing less than a new planet.

It was an extraordinary keenness of eye and judgment that enabled Herschel to distinguish the planet from the nearby stars by its appearance alone. Other observers, while measuring the positions of neighboring stars, had 17 times measured the position of Uranus and had noticed no unusual aspect. Some of the great contemporary astronomers had difficulty in identifying the planet even after they had been informed of its exact position on the sky.

Uranus did not become the planet's official name for several years. It first bore the title "Georgium Sidus" (Herschel's appellation in honor of King George the Third), and was also called "Herschel" for its discoverer. The present name was finally adopted, in conformity with the naming of the other planets.

In spite of Uranus' slow motion (with a period of 84 years), its orbit could be well determined within a relatively short time after discovery because of the 17 inadvertent observations made before Herschel noticed the disk. The first observation, made in 1690, was earlier by nearly a complete revolution of Uranus. The orbit calculators found some difficulty in reconciling all the observations, but the chance for errors in the observations, or for deviations because of the perturbations by other planets, seemed great enough to account for the discrepancies. However, when Uranus began to deviate appreciably from its computed path, even after careful allowance had been made for the perturbations by Jupiter and Saturn, several astronomers began to suspect that an unknown planet might be disturbing the motion of Uranus.

In the second and third decades of the nineteenth century the deviations were large enough to arouse suspicion, but the mathematical difficulties in predicting the position of the unknown planet seemed insurmountable at that time. By the year 1845 Uranus had moved out of place by the "intolerable quantity" of 2 minutes of arc, an angle barely resolvable by the naked eye. Urban Jean Joseph Leverrier (1811–1877), the great

French astronomer, showed in 1846 that no possible orbit for Uranus could reconcile all the observations within their reasonable errors. He concluded that the deviations could be explained only by the hypothesis of an unknown massive planet beyond the orbit of Uranus.

Later in the year 1846 Leverrier completed his calculations for the position of the hypothetical planet, and was so confident of his analysis that he dared to predict its position and that it would show a recognizable disk. He sent his predictions to the young German astronomer J. G. Galle (1812–1910), who discovered the actual planet *on the same night* that he received the prediction. Neptune's position on the sky lay within 1°—less than two Moon diameters—of the position forecast by Leverrier. Galle's immediate success was due to his access in Berlin to a new star chart of the appropriate sky region. A quick telescopic survey showed the new object where no star had been seen before. A close inspection verified the existence of a disk, too small to be distinguished easily.

This remarkable discovery of a new planet by means of mathematical deduction is a landmark in the history of astronomy. Like many great discoveries, it must be credited to more than one man. While Leverrier had been making his brilliant calculations, a young and unknown English mathematician, J. C. Adams (1819–1892), arrived independently at the same result by a somewhat different method. Adams's calculations had, indeed, been completed some eight months before Leverrier's, but unhappy circumstances prevented the English observers from anticipating Galle's discovery. The Berlin star chart that had so materially assisted Galle was not then available in England. Hence the astronomer James Challis (1804–1883) at Cambridge half-heartedly began searching for the planet by the arduous method of plotting all the stars in the region, with the intent of reobserving them later in order to detect the planet by its motion. Neptune might also have been found at the Royal Observatory in Greenwich except that the Astronomer Royal, G. B. Airy, held a negative attitude toward theory and simply did not believe that Adams could make a valid prediction.

The entire turn of events was heartbreaking for Adams, who had apparently planned his investigation some years before he had the opportunity to execute it. Posthumously the following note was found among his effects: "1841, July 3. Formed a design, in the beginning of this week, of investigating, as soon as possible after taking my degree, the irregularities in the motion of Uranus, which were as yet unaccounted for: in order to find whether they may be attributed to the action of an undiscovered planet beyond it, and if possible thence to determine the elements of its orbit, etc., approximately, which would probably lead to its discovery."

It is a pleasure to record that both Leverrier and Adams now share equally the honor of having predicted the existence and position of Neptune. Galle, of course, receives his full credit for actually having found it. Like Uranus, Neptune had been mistaken for a star in the course of previous measurements of stellar positions.

The conquest of the solar system by Newton's law of gravitation and by painstaking observation has been continued in the present century. Efforts have culminated in the discovery of Pluto, under circumstances surprisingly similar to those related for Neptune. Again, an early search actually included the new planet but perverse fortune prevented its detection until much later.

At the beginning of this century Percival Lowell (1855–1916; Fig. 29), who founded an observatory at Flagstaff, Arizona, for the purpose of observing the planets, particularly Mars, became actively interested in a possible trans-Neptunian planet. He reinvestigated the orbit of Uranus and concluded that the apparent errors of observation could be materially reduced by the inclusion of perturbations by an unknown planet. His calculations of the orbit and positions of Planet X were not published until 1914, although his search for the planet was begun in 1905. Twenty-four years later, in 1929, a new 13-inch refracting telescope to expedite the search was completed and was installed at the Lowell Observatory.

A young assistant, Clyde Tombaugh, was assigned the task of systematically photographing regions of the sky along the ecliptic. For each region he made two long-exposure photographs, separated in time by 2 or 3 days. Then, in search of the predicted planet, he very carefully compared the resulting photographic plates. Comparisons were made by means of a *blink comparator,* a double-microscope apparatus that enables the

Fig. 29. Percival Lowell, whose prediction and enthusiasm led eventually to the discovery of Pluto.

observer to inspect the same area of the sky on two plates alternately (Fig. 30). Any object that has moved on the sky during the interval between the two exposures appears to jump back and forth among the stars, which appear to remain fixed (Fig. 31).

Fig. 30. Clyde W. Tombaugh at the blink comparator where he sat for 7000 hours in his planetary search. (Photograph by the Lowell Observatory.)

On March 12, 1930, less than a year after the institution of its new observing program, the Lowell Observatory telegraphed astronomical observatories the following announcement: "Systematic search begun years ago supplementing Lowell's investigations for Trans Neptunian planet has revealed object which

Fig. 31. Discovery photographs of Pluto: (*left*) January 23, 1930; (*right*) January 29, 1930. (Photographs by C. W. Tombaugh with the 13-inch Lawrence Lowell Refractor of the Lowell Observatory.)

since seven weeks has in rate of motion and path consistently conformed to Trans Neptunian body at approximate distance he assigned. Fifteenth magnitude. Position March twelve days three hours GMT was seven seconds West from Delta Geminorum, agreeing with Lowell's predicted longitude."

The astronomical world soon unanimously adopted the name Pluto as appropriate to this planet, which moves in the outer regions of darkness. The first two letters of the name are, moreover, the initials of Percival Lowell, who had died in 1916, only two years after the publication of his detailed prediction.

Subsequent orbits, based on prediscovery photographs of the new planet, show that it moves about the Sun with a period of 249.9 years, in an orbit inclined 17° to the mean plane of the other planets. At perihelion the orbit passes within that of Neptune, but because of the high inclination the two bodies cannot collide.

Only the perversity of chance kept the discovery of Pluto from being made by the Mount Wilson astronomers in 1919. At that time Milton Humason, at the request of William H. Pickering (1858–1938), who had independently made calculations of the assumed planet's position, photographed the regions around the predicted position and actually registered the planet on some of the plates. Pluto's image on one of the two best plates, however, fell directly upon a small flaw in the emulsion —at first glance it seemed to be a part of the flaw—while on the other plate the image was partly superimposed upon that of a star! Even in 1930, when the 1919 position was rather well known from the orbit, it was difficult to identify the images that had been produced by Pluto eleven years before.

Pluto's mass is not yet certainly determined but it cannot be massive enough to have produced the deviations in planetary motion on which its prediction was based. A few stellar occultations by Pluto set limits on its diameter to Moon-like dimensions. The probable existence of a near satellite limits its mass and density to values consistent with its dimensions. The basis for its prediction are undoubtedly systematic errors of an arcsecond or two in the early observations of Neptune and Uranus. Nevertheless, the discovery, following the relentless search for the planet, represents a crowning achievement.

Tombaugh has continued the Lowell Observatory search,

covering the entire sky, but finds that there exist no more planets within the discovery range of the 13-inch telescope. If other planets exist, they must be considerably fainter than Pluto, which means that they must be either much farther away or smaller. Some astronomers doubt that there are more sizable planets to be found. Tiny ones, of asteroid dimensions, may well be numerous in the outer solar system.

4

Weights and Measures

Weighing the planets and finding the distances between them are naturally most important in learning about their true character. Only from a knowledge of the masses can we begin to ascertain the real structure of the individual bodies. Furthermore, the landing of space vehicles on planets and satellites demands the utmost precision in distances, dimensions, and masses. Thus to increase our knowledge, we place demands on the accumulated fund of knowledge.

The Distance to the Sun

The stars, apparently bright points scattered all around the sky, form a magnificent reference system against which we can measure the directions of moving bodies in the solar system. As we noted in Chapter 2, the stars are so distant that our motion does not affect their directions perceptibly, except in a few cases. Thus, by measuring planetary and solar directions precisely with respect to the stars, we determine them, with respect to a well-defined reference system.

With accurate observations and calculations now spread over hundreds of years, we can apply Newton's law of universal gravitation and calculate all the relative directions and distances to the Sun and planets with a precision of about one part in a million. But all these accurate distances are determined in terms of the astronomical unit, half the length of the Earth's ellipse about the Sun, not in terms of meters or kilometers. For purposes of prediction, the arbitrary nature of this unit of distance makes practically no difference, but no scientist relishes a measuring rod whose length is unknown. Also, in maneuvering space vehicles we need to know the actual distances.

Until the last few decades the astronomer had only one type of measure at his disposal—the direction of the object. Radio added a new type of measure, both for calibration of the astronomical unit and for direct distance measurements to space craft, moons or planets. In radar, a transmitter with an antenna, sometimes huge, sends out short pulses of perhaps centimeter-wavelength radiation. When these pulses bounce back from a target surface, say the Moon, the time lag gives the distance. The speed of light is an absolute constant, 299,792.5 kilometers per second in the vacuum of space. Small corrections are applied for a slower speed and refraction through the atmosphere. The accuracy obtainable is typically less than a meter, down to a few centimeters.

Variations on the radar technique include a special reflector at the target or an active *transponder;* the latter receives the pulses and immediately transmits back corresponding pulses. Continuous waves or pulses controlled or counted by precision atomic clocks at the source or target can measure changes in distance, giving the velocity. Electronic computers can count the waves or pulses and add them up to provide highly precise distance measures.

The *laser* or *lidar* utilizes the radar principle but at the much shorter wavelengths of light. The light pulses thus permit even greater accuracy than radar in vacuum but are more disturbed by the atmosphere. Laser return power is immensely improved by *corner reflectors* placed on targets such as the Moon or artificial satellites. A corner reflector is literally a square corner made of reflecting surfaces for light or for radio waves. A ball, a light beam, or a radio wave all reflect from a corner precisely back toward the source. In fact, the corner reflectors for lasers must be made with less than the best optical surfaces. Otherwise the returning beam is too narrow to strike the source. A telescope beaming a laser at an Apollo corner reflector on the Moon may move 500 meters or more in the $2\frac{1}{2}$ seconds of pulse travel time. Too narrow a returning beam would miss the telescope.

In calibrating the astronomical unit by classical means we are faced with the fact that the largest available yardstick is the Earth itself; its dimensions are now known rather accurately. But the radius of the Earth is less than 1/20,000 of the astronomical unit; from the Sun it subtends an angle of only 8.8 seconds of arc, the *geocentric parallax* of the Sun (see Fig. 32). Although the Sun's parallax can be measured by simultaneous observations at two widely separated stations, the angle is so small that the percentage error becomes large; hence no precise determination of the length of the astronomical unit can be made by measurements of the Sun.

A better method is to measure in kilometers the distance to some body that comes close to the Earth. Since the body's distance in astronomical units at any time is determinable from many observations and calculations, we can compare the two values to find the number of kilometers in an astronomical unit. The Moon, however, will not serve for this purpose because its distance cannot be calculated in astronomical units without introducing the Earth's mass. As Newton found, the Moon's motion is primarily a measure of the Earth's gravity. Mars, under the most favorable conditions, comes within only 55,700,000 kilometers of the Earth. Venus comes closer, 42,000,000 kilo-

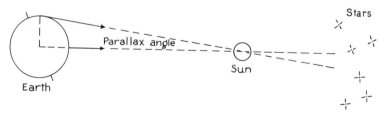

Fig. 32. The solar parallax is the angle subtended by the Earth's radius as seen from the Sun. The geocentric parallax of any celestial object is the corresponding angle from that object. See Fig. 50 for stellar parallax.

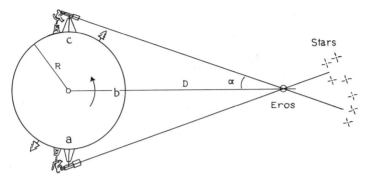

Fig. 33. Measuring the distance to Eros. The observer simply allows the Earth to move him from *a* to *c* in order that he may measure the parallax angle, α, at Eros and calculate the distance, *D*.

meters, but at that time it is nearly in the direction of the Sun and cannot be well measured except by radar (refer to Fig. 4). In the classical astronomical method, however, the best object for the purpose turns out to be an asteroid, Eros.

It is a real pleasure to find a use for an astroid because asteroids are generally more of a nuisance than a help in astronomy. Walter Baade (1893–1960) of Mount Wilson and Palomar Observatories once called them "the vermin of the skies." When Eros comes within 14,000,000 miles of the Earth, it gives a parallax of about seven times that of the Sun, or over a minute of arc. Observers, instead of choosing different parts of the Earth for their observations, can work independently. Each simply waits for the Earth to turn (Fig. 33) and photographs the asteroid in the evening (*a*), at midnight (*b*), and in the morning (*c*). The position of Eros among the stars changes because of the difference in the position of the observer with respect to the Earth. Knowing the instant of each photographic exposure and his exact position on the Earth, the observer can calculate the distance, *D*, to Eros just as accurately as though simultaneous observations were made at different stations on the Earth.

The new era in measuring astronomical distances began in 1959 when Price and his colleagues at the Massachusetts Institute of Technology pioneered in bouncing radar echoes from Venus (Fig. 34). With the addition of laser reflections from the Apollo corner reflectors, the Moon's distance is now known to centimeter accuracy. Correspondingly, as a result of the space

Fig. 34. The Millstone radar antenna, 26 meters in diameter, that first bounced radio waves off Venus. (Photograph by the Massachusetts Institute of Technology Lincoln Laboratory.)

program we can now calculate the astronomical unit to about the percentage accuracy of the velocity of light, better than one part in a million. The measurement of the solar parallax, or Earth's equatorial radius seen at one astronomical unit, is 8.79418 arcsecond and is continuously being improved. The aiming and guidance of space probes has become a technology of precision.

Weighing the Earth

The first step in weighing the planets is to weigh the Earth. The ancient Greek mathematician Archimedes (of Syracuse, 287–212 B.C.) said that if he had a place to stand he could move the Earth. He could equally well have weighed it by observing how easily it moved when he pulled the levers. We are actually interested, not in the weight of the Earth, but in its *mass*. The weight of a body is only a measure of the Earth's attraction for it, while its mass represents the quantity of matter that it contains. One of Newton's great discoveries was that mass and weight are proportional. If we go back to Newton's laws of motion we find that mass is the measure of the force necessary to change the motion of a body by a certain amount. It takes more force to accelerate a 10-ton truck than a baby carriage at a given rate because of the difference in mass. In empty space away from attracting bodies, neither the truck nor the baby carriage would have any weight but their masses would remain unchanged.

We know just how much force the Earth exerts on a unit mass through gravitation. This force is the surface gravity, which holds us down and enables us to *weigh* things. Since gravity is proportional to the Earth's mass, the only unknown is the constant of attraction between two masses, that is, the *constant of gravity.*

One obsolete method of finding this constant is by measuring the attraction of a mountain on a plumb line. As in Fig. 35, the plumb line does not point straight up but points away from the mountain, because the bob is attracted toward it. We measure the force exerted on the bob by the mountain and estimate the mass of the mountain by measuring its size and composition.

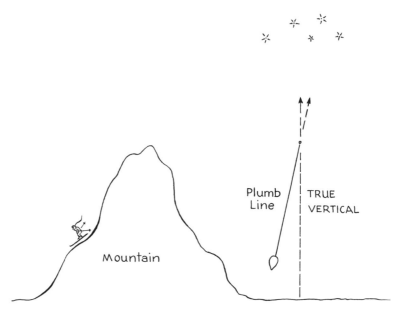

Fig. 35. A mountain attracts a plumb bob. The direction of the plumb line deviates from the vertical, to provide a measure of the mountain's gravitational force.

Since the distance to the mountain is measurable, we can calculate the constant of gravity and hence the mass of the Earth.

The mountain method is extremely poor compared with laboratory methods. Exceedingly delicate instruments can measure the attraction of a large ball of lead upon a smaller ball directly, giving a value for the constant of gravity. Since the weight of the small ball measures the Earth's attraction for it, the mass of the Earth is then determined in terms of the mass of the large ball by means of the inverse-square law of the attractions. If the Earth could be put on scales at the surface, it would weigh 5,976,000,000,000,000,000,000 metric tons.

The force of gravitation is an exceedingly trivial force, becoming appreciable only when huge quantities of matter are involved. Suppose a ball were made of all the gold that has been mined in the world, say 90,000 metric tons; the ball would be 20.7 meters in diameter. If it were placed in space, away from other attracting forces, a 90-kilogram person sitting on it would weigh the equivalent of one-half gram on the Earth. A cricket could easily lift him and a frog could kick him completely away from the ball of gold. Since humans are not usually so easily diverted from gold, we may conclude, in the manner of Aesop, that the force of avarice greatly exceeds that of gravitation.

The Massive Sun

Knowing the mass of the Earth, we can now calculate the mass of the Sun. The Earth continually falls toward the Sun, away from the straight line that it would follow if there were no gravitational attraction. The amount of fall is about 3 millimeters every second, in which time the Earth moves forward about 30 kilometers. The mass of the Sun required to make the Earth fall at this rate is 332,800 times the mass of the Earth, or some 1.989×10^{27} (1,989,000,000,000,000,000,000,000,000) metric tons.

With this knowledge of the mass of the Sun we may deduce some interesting information about its constitution. The average density is only 1.41 times that of water, while the Earth's density is 5.5, equivalent to a mixture of rock and metals. At the surface of the Sun the force of gravity is 28 times as great as on the Earth. A 90-kilogram person would weigh nearly 3 tons there, only to evaporate instantly at the temperature of 5,500°C. From only three known quantities—mass, diameter, and surface temperature—it is possible to prove that the Sun is a *gas* throughout. The temperature at the center must be about 15,-500,000°C, with a density of 160 times that of water, to provide enough pressure to keep the outer gaseous layers from collapsing. No element can be a solid or a liquid in any part of the Sun; even tungsten, used in electric-light filaments, would evaporate on the surface, which is relatively cool.

A Planet with a Satellite

For a planet with a satellite the method of determining the mass is like that used in finding the Sun's mass. The attraction of the planet must always exactly equal the centripetal force, which measures the rate at which the satellite falls toward the planet to remain in its orbit. With a knowledge of this attraction, the distance of the satellite, and the constant of gravitation, we calculate the mass of the planet. We knew the mass of Neptune,

4,500 million kilometers away, as accurately as we knew the mass of the Moon, distant only 384,401 kilometers, at least before radar and lasers.

Weighing the Moon

The mass of a satellite is difficult to determine because it is generally so small compared with the mass of the primary planet. The effect of the Earth on the Moon's motion is easily measured, but the Moon is so small in mass that it affects the Earth's motion only slightly. The center of the Earth moves about their common center of gravity in a very small orbit, identical in shape with that of the Moon. If Earth and Moon could be joined by a weightless rod and the rod balanced on a knife-edge under a constant gravity, the knife-edge could support the rod at their center of gravity (Fig. 36). It is this point that moves in a smooth elliptic orbit about the Sun. By carefully observing the distance of the center of gravity from the center of the Earth, we can measure the mass of the Moon (Fig. 37). The ratio of the radius of the little orbit of the Earth's center to that of the larger orbit of the Moon is the ratio of the Moon's mass to that of the Earth. Subsatellites moving about the Moon can, of course, provide a better measure of its mass, as can radar and lasers.

The center of gravity is about 4730 kilometers from the center of the Earth, so that the mass of the Moon is only 1/81.30 (or 4730/384400) times the mass of the Earth. With such a small mass, only 73,500,000,000,000,000,000 metric tons, the Moon is as tiny a part of the entire solar system as a drop of water in a 50-gallon barrel, or as the proverbial fly on a cartwheel. The Earth is merely 81 times as important—except to us.

Fig. 36. Center of gravity is the point of balance. The figure is accurately drawn for two balls of equal density connected by a weightless rod.

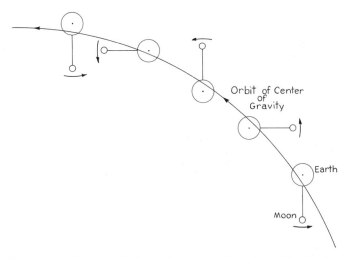

Fig. 37. Motion of the Earth about the center of gravity of the Earth and the Moon. The center of gravity moves in a smooth ellipse about the Sun. The dimensions are exaggerated, but the relative positions of the Earth's center are correct.

Other Satellites

Only a few of the larger satellites of Jupiter and Saturn produce sufficient gravitational effects for their masses to be determined. Their mutual perturbations measure their masses with respect to their primaries. Close flying spacecraft are also perturbed, leading to improved measurements of the masses of satellites. Jupiter's four bright satellites and Titan of Saturn's system are comparable to the Moon or somewhat more massive. Jupiter's Ganymede and perhaps Neptune's Triton have about twice the mass of the Moon, although Triton's mass is not yet well determined.

Radii, masses, densities, and other data concerning the satellites are listed in Appendix 3. The relative masses of the satellites are represented by spheres of equal density in Fig. 38. For those whose masses are not known, the measured diameters are given instead, and the satellites are designated by question marks. Diameters that were estimated from brightness alone are similarly designated. The range in mass is so great that it was necessaary to magnify some of the diameters by a factor of 10 or 100 to make them visible on the diagram.

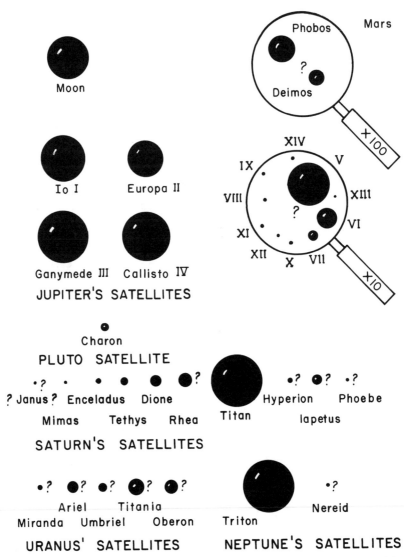

Fig. 38. The masses of the satellites in the solar system are depicted by spheres of equal density. Uncertain masses are designated by a question mark. The magnifications are in terms of diameter. The satellites are placed in the order of their distances from their respective primaries. See Appendix 3 for other data about the satellites.

The Moon has a density of 3.3 times that of water as though it were made of ordinary rock. Among the larger satellites only Jupiter's remarkable Io is denser than the Moon, 3.5 times water. The densities of Jupiter's great satellites fall with increasing distance from the planet: Europa, 3.0; Ganymede, 1.93; and Callisto, 1.79. Ganymede and Callisto, Titan of Saturn, and Triton of Neptune about match Mercury in diameter but are far less dense. Satellites with densities much below three times water must contain a considerable fraction of water or ice because few types of minerals have such low densities.

The Masses of the Planets

The masses of planets without satellites have been calculated classically only from their observed perturbations on the motions of the other planets, by a method similar to the one used in predicting the positions of Neptune and Pluto. Venus, for example, approaches the Earth, Mars, and Mercury fairly closely and produces perturbations in their motions. Similarly, the motion of Venus is perturbed by the attractions of the Earth, Mars, and Mercury. The numerical calculation of the masses of Venus and Mercury from observations of their motions is exceedingly complicated. Artificial satellites or space probes about these planets vastly improve the values of their masses and, of course, our knowledge of their shapes and surface topography. Furthermore, as we shall see, the orbiting spacecraft can tell us about the internal structure of planets.

Mercury turns out to be extremely dense, 5.44 times water, intermediate between Venus (5.25) and Earth, the densest planet (5.52). Even though Mercury is only 1.4 times the Moon in diameter, its density is 1.62 times greater, indicating a highly important difference in composition.

Pluto's mass and density will remain unknown until its satellite has been fully confirmed and well observed. As noted earlier, Pluto is probably about Moon sized and contains a good fraction of ice.

Figure 39 shows the masses of the planets as balls of equal density. For comparison with Fig. 38, we recall that Mercury surpasses the Moon and other great satellites in mass if not always in diameter. Therefore the sequence of masses from the

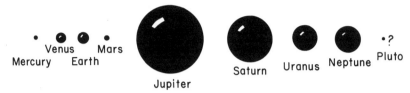

Fig. 39. The masses of the planets, represented by spheres of equal density. The Sun's diameter, on this scale, would exceed Jupiter's by a factor of ten, and its sphere would cover the entire page.

least satellite, Phobos, to the greatest planet, Jupiter, is fairly uniform. We might extend the sequence to the smaller bodies, such as the asteroids and meteors, without loss of uniformity, but not in the other direction. The step in diameter from Jupiter to the Sun is a factor of ten, because the Sun is a thousand times more massive.

It would unduly burden this story to describe all of the other ingenious methods that have been used in calculating or estimating the masses of planets, satellites, comets, or asteroids. Whenever two bodies, including spacecraft, are observed to come near enough together for one to perturb the motion of the other, additional information about the masses can be determined. If close approaches occur without any observed changes in motion, an *upper limit* to the masses can be deduced. When Brook's comet, in 1886, came within the orbits of Jupiter's inner satellites, the comet's period of revolution about the Sun was changed from 29 years to 7 years, yet no change was observable in the motions of the satellites. The comet consequently must have possessed less than 0.0001 of the Earth's mass, or it would have produced measurable perturbations.

In concluding these chapters that involve Newton's law of gravitation we note that the whole foundation of astronomy rests on applications of the law as do the motions of man-made vehicles in space. Outside the solar system in the far reaches of the universe, the law is still the key to the solution of many of the most important problems. Almost no calculations of mass can be made without using the property of attraction. The correction required by Einstein's general relativity is just appreciable in the case of Mercury's rapid motion under the Sun's great attraction. Solely in this instance are these corrections great enough to be detectable in present-day observations of the motions of planetary bodies. The increasing precision, however, can be expected to reveal other instances in the future. Since there can be no absolute truth, each new conclusion leading on to the possibility of more general ones, we can well admire the simplicity and perfection of Newton's law which applies so exactly. To make progress, however, the scientist must search for tiny imperfections in a law or theory that seems to be perfect.

5

The Earth

Our Earth seems so large, so substantial, and so much with us that we tend to forget the minor position it occupies in the solar family of planets. Only by a small margin is it the largest of the other terrestrial planets. True, it does possess a moderately thick atmosphere that overlies a thin patchy layer of water and it does have a noble satellite of about one-fourth its diameter. These qualifications of the Earth, however, are hardly sufficient to bolster our cosmic egotism. But, small as is the Earth astronomically, it is our best-known planet and therefore deserves and has received careful study.

Before the days of artificial satellites the dark part of the new Moon served as our best approximation to a mirror in space for studying the Earth (see Fig. 40). Near the new phase, when the Moon lies almost in a line with the Sun, the light reflected from the Earth illuminates the otherwise unlighted black hemisphere. Measures of the earthshine on the Moon indicate that the Earth is a good reflector of light, as are the other planets

Fig. 40. Earthshine on the Moon in an early phase. (Photograph by the Yerkes Observatory.)

with atmospheres. Because of the variable cloud cover, however, the fraction of sunlight reflected from the Earth, its *albedo*, varies over a range of about a factor of 2, averaging 0.35.

The remarkable photographs taken by rockets (Fig. 41), by the U.S. Weather Satellite TIROS, and by several other U.S. spacecraft show the Earth from outside (see Fig. 42). As expected, the clouds form the most conspicuous features at any time (Fig. 43). The great extent of some of the cloud banks, however, proved a surprise to the meteorologists; great waves of weather on the Earth stand out in the cloud pictures. Meteorological satellites, with their wealth of data on the worldwide state of the atmosphere, are changing the art of weather forecasting into a science.

A hypothetical visitor at some distance in space, if equipped with a good telescope, could eventually distinguish the continents from the oceans and identify the polar caps. During the winter season in our Northern Hemisphere he would see the north polar cap covering an enormous area, some 50° from the Pole, while in summer the area would shrink to only a few degrees of latitude. The lower border would always be very irregular, particularly where it is broken by the oceans. The south polar cap would change very much less because of the scarcity of land. The seasonal changes from green to brown to black and

Fig. 41. Southwestern North America as photographed by a U.S. Navy Aerobee rocket from an altitude of 160 km. (Official U.S. Navy photograph.)

white in the temperate zones could probably be recognized and their causes explained by a clever observer.

One peculiarity that we cannot observe on any other planet could be seen by our hypothetical astronomer outside the Earth. He would be able to observe the *direct reflection* of the Sun from our oceans. The phenomenon might be a great surprise for an astronomer who had never encountered large bodies of water. He might very well attribute the bright reflection area to a smooth crystalline surface on the Earth, as the early astronomers visualized the Moon to be a perfect crystal sphere.

One observation about the planet Earth, as recorded by an outside astronomer in his book of facts, would be that the axis of rotation is not perpendicular to the ecliptic, the plane of revolution about the Sun. By long and careful measures he would conclude that the equator is tipped 23.5° from the plane of the ecliptic. This *obliquity of the ecliptic* might enable him to account for the seasonal changes of color in the temperate zones and the variation in the sizes of the polar caps.

He would conclude that the direction of the axis remains fixed in space as the Earth moves in its orbit about the Sun, as in Fig. 44. When the North Pole was tipped toward the Sun (*a*), that hemisphere would be more illuminated by the Sun's rays. The Pole would be lighted continuously and the length of the day would be greater everywhere north of the equator. In addition, the light would fall on the surface at an angle more nearly perpendicular so that a given area would receive more heat and light during the daylight (Fig. 45), and would have more hours of sunlight.

A quarter of the period (3 months) later, the Earth would be

Fig. 42. The Earth as seen from space: a view of Central Africa from spacecraft Mercury Atlas 4, showing Lake Rudolf in Kenya. (Courtesy of the National Aeronautics and Space Administration.)

Fig. 43. The Earth as seen by the U.S. applications Satellite 1 on December 9, 1966. Baja California shows in the upper right center and the solar reflection as a haze to left and below center. (Courtesy of the National Aeronautics and Space Administration.)

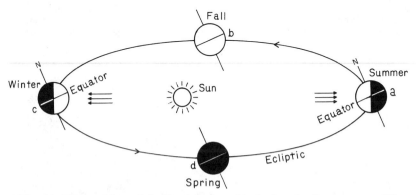

Fig. 44. The position of the Earth in its orbit during the four seasons. Winter and summer are both moderated in the Northern Hemisphere and accentuated in the Southern Hemisphere by the eccentric position of the Sun. The seasons, labeled for the North, are reversed for the South.

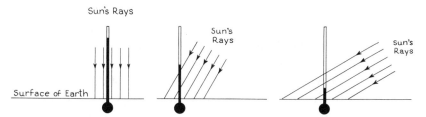

Fig. 45. The Sun heats a surface more effectively when the rays fall vertically, rather than obliquely.

in position *b*, Fig. 44, and everywhere the hours of daylight would equal those of darkness. During the next half period (6 months) the Southern Hemisphere would gain in heat while the Northern Hemisphere would lose, the North Pole being completely dark during that time. If our outside astronomer possessed any ingenuity at all (and of course he must, being an astronomer!), he would be able to explain completely the observed changes in color in the two hemispheres, and also the peculiarities in the changes of the polar caps. He might express himself as follows: "Clearly on the planet Earth there must be complicated chemical or physical reactions that are directly activated by solar heat. Some regions, the dark-blue areas that constitute most of the planet's surface, are affected only by very great variations in heat while other regions, those that turn green as the temperature rises, are affected by much smaller changes. The permanent polar caps are probably similar to those regions that turn green with increase of temperature, but are never heated sufficiently for the reaction to occur." Our learned friend from outside might continue, "We must conclude, therefore, that the more stable, dark-blue areas are very good conductors of heat as compared with those unstable areas which are so affected by slight changes . . ."

The time of highest temperature would be observed not to occur at the time of greatest length of day and of maximum sunlight on the surface. In the North Temperate Zone the maximum sunlight falls near June 21 (point *a* in Fig. 44), but midsummer, the time of highest temperature, comes actually late in July or near the first of August. The other seasons are correspondingly late. The seasons *lag* because the surface of the Earth (only the upper few meters and the atmosphere) becomes warmer as the amount of heat received from the Sun increases. The temperature continues to rise as long as the heat is strong, even though it is beginning to wane, until the rate of gain equals the rate of loss. Similarly, the coldest winter weather comes a month or more after December 21, the shortest day of the year.

It is interesting to note that the Earth is at perihelion, nearest to the Sun, during midwinter in the Northern Hemisphere and at aphelion, farthest away, during midsummer. The effect is to moderate the seasons slightly in the Northern Hemisphere, but to increase the range in temperature slightly in the Southern Hemisphere. In fact, however, the seasonal temperature changes in the Southern Hemisphere are smaller than in the northern, for a different reason. The oceans tend to control and moderate temperature changes in the atmosphere; the ratio of water area to land area is much greater in the Southern than in the Northern Hemisphere.

The Earth is generally called a sphere but actually is not a perfect one. Careful measures show that the diameter at the equator is 43 kilometers (one part in 298) greater than the diameter at the poles; the equatorial cross section is bulged outward. This deformation does not arise by chance; the Earth was not cast in that shape to remain forever. The internal gravitation is great enough to draw the material of the Earth into an almost perfect sphere, were there no rotation. The rotation in 24 hours, however, produces a centrifugal force that enlarges the equatorial diameter at the expense of the polar diameter, to produce the observed equatorial bulge. If the underlying material of the Earth were too rigid to take the form prescribed by the rotation, the water in the oceans would flow to the equator to compensate the centrifugal force (Fig. 46a). Since the oceans are not particularly deeper there than elsewhere on the globe, we must conclude that the "solid" Earth adjusts itself to the force (Fig. 46b).

The presence of an equatorial bulge on the Earth, besides leading to such paradoxes as "The Mississippi River runs uphill," has one effect of great importance to the astronomer. This effect is called the *precession of the equinoxes,* observed in antiquity and explained by Newton. The term "precession" describes the fact that the Earth's axis does not remain fixed in direction over long intervals of time, but moves slowly around with a period of

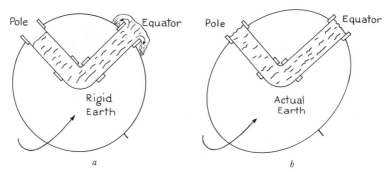

Fig. 46. (a) On a rigid spherical Earth in rotation, water would spill out of a pipe running from the pole to the equator. Although no such pipe exists, the oceans, correspondingly, would flow to the equator. (b) The actual Earth in rotation bulges at the equator, and hence the water level is uniform between the pole and the equator.

about 26,000 years. The angle between the equator and the ecliptic does not change essentially, although the axis twists around like the axis of a spinning top. The analogy (Fig. 47) is almost perfect, for the Earth really acts as a huge top.

The axis of the top is the polar axis of the Earth, the main body of the top is the Earth, and the rim of the top is the equatorial bulge. Because of the obliquity of the ecliptic, the bulge is

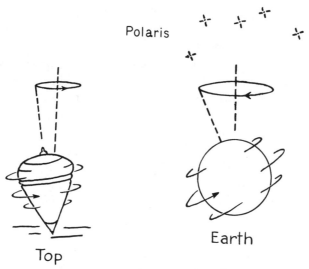

Fig. 47. A top and the Earth. Both precess because of forces that act to change the direction of their axes of rotation.

always being attracted by the Moon, the Sun, and the planets, which try to turn the bulge, and therefore the equator, into the plane of the ecliptic. In the case of the top, the action of gravity is the reverse, tending to overturn the axis of rotation. In neither case does the overturning force succeed in causing the spinning body to tip over. Instead, the angle between the spin and the force remains the same but the axis precesses around as shown in Fig. 47. The peculiar property of a spinning body to resist a force applied to the axis is exemplified in the gyroscope, an instrument whose most important use is in gyrocompasses, in stabilizers, and in attitude-control units for ships, airplanes, rockets, and space vehicles. Gyroscopic control units are now amazingly precise, capable of being used as clocks as the Earth turns about them. They can hold orientation to the order of one part in a million.

The precession of the equinoxes ceases to be a purely academic problem when we look into the complications that it produces in calendar making. Back in Fig. 44 we see that the time of the seasons will depend on the direction of the Earth's axis. When the Earth and the Sun are on the line of the *equinoxes* (*b* or *d*), the line along which the planes of the equator and ecliptic intersect, the season will be either fall or spring. The precession of the equinoxes is a westward motion of the equinoxes as measured with respect to the stars, that is, clockwise when one looks down on the North Pole. If the year were defined as one revolution of the Earth about the Sun as measured with respect to the stars, the seasons would soon begin to get out of step with the months and within a few thousand years would be entirely changed. To avoid such a difficulty the calendar year, or *tropical* year, is measured from the time that the Sun is in the direction of the *vernal* (spring) equinox until it has returned there again. This tropical year keeps the calendar in step with the seasons but is shorter than the true, *sidereal,* year by about 20 minutes. The leap-year troubles of calendar making arise from the 5 hours 48 minutes 45.6 seconds that the tropical year exceeds 365 days in length.

Ancient astronomers noticed that the position of the stars in the sky at the same season shifted slowly with time, and also noticed the more obvious phenomenon that the North Pole of the Earth was shifting among the stars. Polaris, our present pole star, is only temporarily useful as such (see Fig. 200, star chart at end of book), though the motion of the pole in a lifetime is negligible. At the time when the Pyramids of Egypt were constructed, the pole star was α Draconis, some 25° from Polaris. The Southern Cross (Crux, a southern constellation) could then have been observed from most of the land that now makes up the United States. The pole moves in a small circle of radius 23.5° in the mythical "Year of the Gods." To these deities a man's life seems like a day. The period of precession is close to 70×365 years.

Since the Moon provides the major force on the Earth's equatorial bulge to produce the precession of the equinoxes, the bulge must, by Newton's third law of motion, affect the Moon's motion. It does; the major effect is to swing the plane of the Moon's orbit around (westward or backward) much in the manner of precession but in a period of only about 19 years. On nearby artificial Earth satellites the equatorial bulge has a profound effect, swinging the orbital planes around in weeks instead of years. Even though the computations of satellite motions are greatly complicated by these motions, they lead to some excellent new results. Not only is the oblateness of the Earth much better determined; the satellites show that the Earth is not gravitationally symmetric about the equator. It is slightly pear-shaped with the stem at the North Pole (Fig. 48).

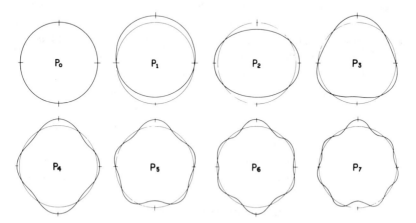

Fig. 48. The pear-shaped Earth is indicated by the diagram P_3, which represents only the third term in an infinite series giving a mathematical expression of the Earth's true shape. Artificial Earth satellites now give values for dozens of terms in this series. The term P_2 represents the Earth's oblateness. The North Pole is up.

The distortion amounts to about 16 meters. Part of the pear shape arises from a pair of opposed gravitational "dimples" on Earth in the Northern Hemisphere, near India and off the west coast of North America. The equivalent depressions amount to about 75 meters and are derived from density variations within the Earth's crust. Possibly they measure the effects of currents within the "solid" Earth (see Chapter 6).

To the astronomer the precession of the equinoxes presents many more serious difficulties than are involved in a mere change of the length of the year. He is forced to make his measures of celestial objects, natural or artificial, with respect to a reference system that does not remain fixed. His plight resembles that of an imaginary map maker who finds that all the continents and islands of the world are moving out (actually they do, at rates of centimeters per year as we shall see later). To state the latitude and longitude of a point on one of them would require a calculation involving the desired instant of time. Similarly in astronomy the starting point for practical measures is the vernal equinox, and the fundamental plane is the equator. Since these primary directions are moving because of precession, every published measure of a star, planet, or satellite must carry with it the date of the equinox and equator to which the measure is referred.

The motions of the Earth cause two additional complications in the problem of recording directions on the sky. One of these, *nutation,* is a small periodic irregularity in the precession, or the motion of the Earth's polar axis among the stars; it is caused by the Moon's peculiar motions and varying attraction on the equatorial bulge. The main effect of nutation is an oscillation in the motion of the Earth's axis over a period of about 19 years. Only the most complicated mathematical theory enables the astronomer to calculate all of the small disturbances that finally add together in producing precession and nutation. Professor F. R. Moulton (1872–1952) said, "No words can give an adequate conception of the intricacy or the beauty of the mathematical theory of nutation."

The second complication in observing the celestial bodies is the effect known as the *aberration* of light. Aberration was first observed and explained in about 1728 by the English Astronomer Royal, James Bradley (1693-1762). The story is one of the many fine examples of an exciting type of scientific research in which one phenomenon is sought and an unexpected one found. To prove that the Earth revolves about the Sun, Bradley attempted to observe the parallactic shift of the stars as the Earth moves about its orbit. He set his telescope very rigidly in a well-built (but disused) chimney in order to watch the daily passage of a certain star. If the Earth really revolved about the Sun, the position of each passage should have changed slightly during the year.

Painstaking observations failed to prove the motion of the Earth directly but did show a displacement out of phase with the one Bradley expected. He finally explained the new effect as due to the combined result of the motion of the Earth and the finite velocity of light. There is a well-known story that the explanation came to Bradley while he was sailing on the Thames River. The vane on the masthead of the boat changed its direction as the boat changed its course although the wind remained steady from one direction. If we imagine the wind as being light from a star, the boat as our moving Earth, and the vane as a telescope pointing in the apparent direction of the star, we can see that the direction of the star will depend upon the motion of the Earth. In Fig. 49, where a telescope and an incoming light ray are shown, one can see that the motion of the telescope while the light is passing through will necessitate that the telescope be tilted forward in order to prevent the light ray from striking the sides of the tube.

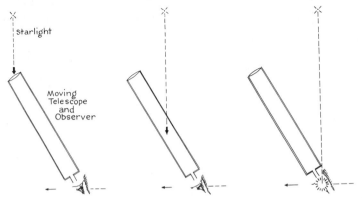

Fig. 49. Aberration of light. A moving observer must tip his telescope forward in order that the moving light ray may pass centrally through the tube.

A good example of aberration is the everyday experience of slanting tracks of raindrops down the side windows of a moving car. Drops that actually fall vertically may leave highly inclined tracks. When light falls upon the Earth from a star, the change in direction by aberration is small, about 20 seconds of arc, but great enough that all observations of celestial bodies must be corrected for it. The velocity of the Earth is only 29.8 kilometers per second, as compared with the velocity of light, 299,792 kilometers per second. The ratio of these two velocities corresponds to the aberration angle (its tangent).

It is noteworthy that though it was Bradley also who discovered nutation, he never did attain his original goal of directly proving the motion of the Earth by the parallactic shifts of the stars. More than 100 years of telescopic improvements were necessary before this result was attained. Bradley's proof of the Earth's revolution about the Sun was nevertheless a good one, despite the fact that it was not the proof he had expected to find.

Not until 1838 was the effect of the Earth's revolution observed directly. Friedrich Wilhelm Bessel (1784–1846) was able to measure a slight shift in the position of the star 61 Cygni as it changed direction when seen from opposite parts of the Earth's orbit. The stellar parallax (Fig. 50), is 0.3 second of arc, that is, the astronomical unit would appear to subtend an angle of three-tenths of a second as seen from 61 Cygni. The nearest star now known is Proxima Centauri for which the parallax is $0''.762$. The distance in kilometers is directly calculated as $(1/0.762) \times 206,265 \times 149,597,000$ or 4.05×10^{13} kilometers. Such a distance is too great to be visualized. It is better expressed in terms of a light-year, the distance that light (moving at 299,792 kilometers per second) travels in a year. Light comes to us from the Moon in 1.3 seconds, from the Sun in 8.5 minutes, but from Proxima Centauri it requires more than *4 years.* Even the bright stars, visible to the naked eye, are much more remote. The most distant star that can be photographed with the 200-inch telescope at Mount Palomar is hundreds of millions of light-years away, while the faintest group of stars, an *external galaxy,* is distant by thousands of millions of light-years. We can scarcely criticize the ancient astronomers for not anticipating these enormous distances.

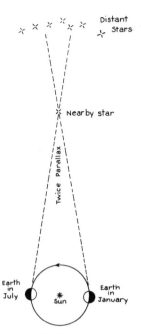

Fig. 50. The parallax of a star is the angle subtended by the radius of the Earth's orbit as seen from the star. The angle in the diagram represents twice the stellar parallax. The parallax can be directly measured for only the nearer stars because it becomes undetectable for the more distant ones.

The motion of the Earth is considered today not as the ponderous movement of a huge mass through space but as the natural movement of a small planet in the curved space about an average star. In the depths of space the Earth counts for little more than a dust speck. To its inhabitants, however, the Earth is home, the Mother Planet, and is justifiably the most important member of the universe. In the next chapter we shall see how good a home it actually provides.

6

The Earth as an Abode for Life

The Earth is taken for granted by the majority of its inhabitants. There is, of course, some grumbling about the bad weather, the poor crops, or the occasional catastrophes, but generally no critical analysis. Such an attitude was once justified by the fact that there was no alternative; having been born on the Earth, we had no choice but to accept what hospitality it offered. With the conquest of space, however, we may now wish to consider whether it might be desirable to move to another planet or live in a spaceship. Let us therefore adopt a broader viewpoint and look at our mother planet with a critical eye. Let us consider the degree of security the Earth offers, the dangers lurking in space, and the unique conditions required for the maintenance of the fragile force we call *life*.

Let us first consider the dangers from without. For the existence of life as we know it, the immediate temperature must, some of the time, rise above the freezing point of water but must never exceed the boiling point. This restriction on the conditions of temperature is more limited than it seems at first glance, because the temperature scale begins at absolute zero, −273°C and rises indefinitely. The highest temperatures directly observed on stars are some hundred thousands of degrees, while the interiors of the stars possess temperatures of many millions of degrees.

The Sun provides the Earth with the necessary heat to maintain its temperature within a suitable range, only about 100 degrees out of millions, and does not raise the temperature too high. Evidence from the past indicates that the Sun has not *greatly* changed its output of heat for some hundreds of millions of years. The energy of the Sun arises not from any burning process but from nuclear fusion, the transformation of hydrogen into helium, in part by a complicated process involving carbon and nitrogen. The Sun's brightness has theoretically increased a few percent in the last few thousand million years and can be expected to increase at a comparable rate, still negligible in a million years. However, a slight change of only a few percent in the Sun's heat would produce violent changes in the climate of the Earth.

Conceivably such changes in the Sun's radiation have been responsible for the great series of ice ages that have recurred every hundred million years or so. At the moment we appear to be within such a geological period, the ice having retreated only temporarily. The ice ages, however, represent no likely threat to life on the Earth, although they can make certain areas of the Earth uninhabitable for long intervals of time.

The Earth's atmosphere is a vital agent in maintaining a suitable temperature. It acts as a blanket to keep the noon temperature from rising too high and the night temperature from falling too low. Exactly as the glass in a greenhouse transmits the visual light of the Sun but prevents the passage outward of the heat or far-infrared light, to maintain a higher temperature than exists outside, so the atmosphere maintains a temperature balance near the surface of the Earth.

We have seen that the enormous atmosphere of Venus holds the solar heat, raising the surface temperature far above the boiling point of water. We are very fortunate that the Earth was spared this fate, because the Earth and Venus are in many other respects similar. On the Moon, where there is no atmosphere,

Fig. 51. Meteors flash across the field of a fixed telescope. (Photograph by J. S. Astapovitsch.)

the midday temperature surpasses the boiling point of water and the night temperature falls to about $-162°C$, much below the melting point of "dry ice." In space, outside the atmosphere of the Earth and well away from it, the temperature in the shade approximates absolute zero. Clearly a heat-regulating atomosphere is requisite to any active form of life in space.

The atmosphere, moreover, is a protecting roof from more than extremes of temperature. It is an invaluable shield from the meteors continually bombarding the Earth from interplanetary space (see Fig. 51). These meteors meet the Earth at velocities up to 72 km/sec. A meteoric particle weighing only 1/1000 gram, moving at this speed, would strike with the same energy as a direct discharge from a 45-caliber pistol fired at point-blank range. Such a particle would be no larger than a fair-sized speck of dust, smaller than an average grain of sand, yet dangerous to a person. Thousands of millions of such particles strike the

Earth's atmosphere daily, constituting faint meteors that can be seen only with a telescope or by radar. The faintest meteors visible to the naked eye are several times as large. In the atmosphere most of these bodies are quickly vaporized by friction with the air.

In space near the Earth's orbit the tiny meteoritic particles scatter sunlight to produce a glow seen near the Sun just before sunrise in the morning or just after sunset in the evening. Because the dust is concentrated toward the ecliptic, the glow is called the *zodiacal light* (Fig. 52).

It is indeed fortunate that we are shielded from the meteors, but even so, some of the more massive ones are able to penetrate to the surface of the Earth and produce damage. The

Fig. 52. The zodiacal light; the thin vertical line is instrumental. (Photograph by D. E. Blackwell and M. F. Ingham from Bolivia at an altitude of 17,100 feet.)

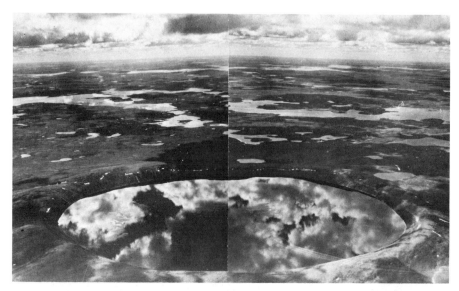

Fig. 53. New Quebec meteorite crater, nearly 5 km in diameter. (Photograph by the Royal Canadian Air Force; courtesy of the Dominion Observatory.)

Great Barringer Meteor Crater in Arizona was formed some 24,000 years ago by the explosion of such a huge body from space. This crater is more than a kilometer in diameter and even now is nearly 200 meters deep despite infilling by erosion. The New Quebec Crater (Fig. 53) in Canada is much larger. Small iron meteorites have been found in abundance around the Barringer crater but no large ones have been discovered either by drilling operations or by radio-detection apparatus. The iron meteoritic body exploded at impact with a force far exceeding that of any known explosives. Only the "shrapnel" and the crater are left to tell us the story.

The Great Siberian Meteor of 1908 detonated so violently that trees were laid flat up to 30 kilometers from the area of impact (Fig. 54). Here the incoming body was almost certainly a low-density comet fragment that broke up in the atmosphere at a height of several kilometers. Tales that this 1908 event, the Tunguska Fall, was a black hole or an exploding nuclear spaceship are refuted by the direct evidence. A large iron meteorite did strike Siberia in 1948, making a number of sizable craters. Also, in 1972 western North America had a near miss by only

Fig. 54. The Great Siberian Meteor devastated the forest over a distance of 30 kilometers from the point of fall. (Photograph by L. A. Kulik.)

some 50 kilometers of a 100-ton or larger mass. Hundreds of vacationers photographed the afternoon fireball in several of the United States and in Canada.

If two large falls and a near miss occurred over land this century, there must have been several more unobserved on the oceans. Had any of these struck a large city, the death toll could have been from ten thousand to a million people, comparable to a nuclear bomb.

Our only protection from such devastating meteors lies in their extreme rarity, but there is always the remote possibility that one of them might be encountered at any time (see Fig. 55). On the average a person is killed by a falling meteorite once in several hundred years.

The elimination of the dinosaurs at the end of the Cretaceous period, sixty-five million years ago, and, indeed, the ending of other geological periods may have resulted from the fall of asteroidal-sized bodies on the Earth. Luis Alvarez and his col-

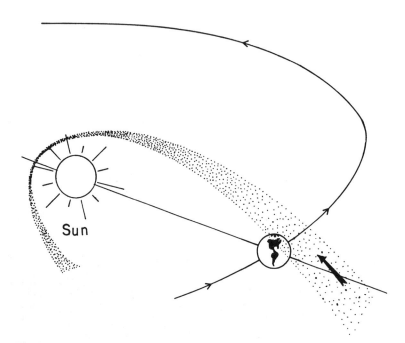

Fig. 55. A stream of cometary debris, moving in an elliptical orbit about the Sun, produces a meteor shower when the Earth passes through it.

leagues at the University of California have shown that iridium is thirty to one hundred sixty times more abundant in the rock stratum marking the end of the Cretaceous than in older and younger strata. Iridium is much depleted in Earth rocks compared with the Sun and meteorites, presumably because it settled to the center of the Earth along with iron. Its concentration in this stratum is strong evidence that an asteroid about ten kilometers in diameter struck the Earth at that time. The mammoth explosion filled the atmosphere with enough dust (more than a thousand cubic kilometers) to block out the Sun for a few years. The lack of photosynthesis halted the food chain, starving all vertebrates weighing more than twenty kilograms and eliminating half of all the genera (classes of species) of living organisms.

Technically, the combined governments of the world could eventually eliminate most of the hazard of huge meteorite falls. A concentrated space watch coupled with a capability for sending rocket propulsion units to the surfaces of potential colliding bodies could deflect their orbits from collision courses. Such an

Fig. 56. Aurora: a rayed band. (Photograph by V. P. Hessler at College, Alaska.)

effort would probably cost as much as the worldwide defense effort but would be a much more successful prevention measure than defense has been against war.

Our atmosphere not only protects us from smaller meteors but also guards us from death-dealing radiations in space. Light in the near ultraviolet causes sunburn but is generally important for health, although not a necessity. Ozone (three atoms of oxygen per molecule) is formed in the atmosphere by the Sun's light; it constitutes a shield from the rays of shorter wavelengths farther into the ultraviolet, where the rays begin to become dangerous to health. The oxygen, nitrogen, and other elements in the atmosphere cut out all of the far-ultraviolet rays below the limit of ozone. Such rays are sometimes used medically to kill bacteria in the air. If they *all* could reach the Earth's surface, it is doubtful that life in any form could exist, at least on land.

Besides the far-ultraviolet rays, many particles that would be dangerous to life are stopped by the atmosphere. These particles, positive and negative in charge, emanate both from the Sun and from space in general. The streams of charged particles that cause the northern lights, or *aurora borealis* (Fig. 56), come from the Sun, while *cosmic rays* come partly from the Sun and partly from distant space. There are other kinds of rays and particles in space that our atmosphere prevents from reaching the Earth. High-energy accelerators can now generate exceedingly dangerous streams of highly energetic particles equivalent to cosmic rays. Cosmic rays striking the Earth and nuclear disintegrations in space produce high-energy radiations such as x-rays and gamma rays, which can be equally lethal.

It is clear that life as we know it requires a very specific set of circumstances for its continued existence on the Earth. The dangers from space specify that the planet must be within a rather definite distance of a star whose light is quite stable over long intervals of time, and the planet must possess an atmosphere capable of regulating temperature and screening out dangerous rays and particles.

No mention has been made of the exact composition required for the atmosphere. Until many more experiments have been made, the limits through which the composition may range without eliminating all possible forms of life are quite uncertain, but probably they are very broad. Oxygen, nitrogen, and carbon dioxide are probably essential components, while water must be available (see Table 1 for the composition of air). Surface water is not indispensable for certain desert plants but water in some form is necessary for all life as we know it. Volcanic gases and chemical combinations with surface rocks probably first determined the composition of the atmosphere. More

TABLE 1. *Composition of air.*

Element	Percent by volume
Nitrogen	78.08
Oxygen	20.95
Argon	0.93
Carbon dioxide	.03
Neon	.0018
Helium	.0005
Methane (CH_4)	.0002
Krypton	.0001
Sulphur dioxide (SO_2)	.0001
Hydrogen	.0005
Water vapor	.2–.4
Traces of other gases and dust	

oxygen and carbon, for example, are now combined in the rocks of the Earth's surface than exist in the air. When life became prevalent, the chemical reactions of the life processes affected the atmospheric composition; plants on land and in the sea as well as silicate rocks now compete for atmospheric carbon dioxide, but the balance is kept fairly stable by the decay of plant life and by the erosion of limestone rocks.

Since World War II high-altitude rockets have been used extensively to sound the upper atmosphere and nearby space (see Fig. 57). Instruments aboard radio their measures down to Earth. They show that the atmosphere has practically a uniform composition to a height of about 90 kilometers. Winds of 160 kilometers per hour and more are observed in the meteor trains (see Fig. 58) and in the strange *noctilucent clouds* seen at the edge of the Arctic circle at altitudes of about 80 kilometers. Such strong winds mix the air sufficiently to prevent the light gases, such as hydrogen, from diffusing appreciably until above an altitude of about 90 kilometers. There are now several methods of measuring the density and temperature of the air at great altitudes. The methods are complicated, depending upon speeds of sound waves from explosions, upon the resistance of the air to meteors, upon the reflections of radio waves, upon rocket-borne laboratories, and, above 160 kilometers, upon the rate at which the resistance of the air brings down artificial satellites.

At low altitudes, up to about 30 kilometers, the temperature

Fig. 57. This Aerobee sounding rocket can carry 100 pounds of measuring instruments well above an altitude of 160 km. (Official U.S. Navy photograph.)

is measured by sending up small balloons with light meteorological equipment. These balloons carry tiny radio transmitters that send down messages of temperature as well as pressure and other characteristics, while their heights are being observed with telescopes or radars. The best estimates of temperatures at

Fig. 58. An enduring meteor train, photographed at 10-second intervals, is rapidly distorted by high-altitude winds. (Photograph by the Harvard Observatory.)

various levels are shown in Fig. 59, together with some of the phenomena that appear at these levels.

The air temperature drops with altitude for a few kilometers in the range of most of our major clouds, where there is good vertical circulation and air cools as it expands on rising. With increasing altitude solar radiation causes a small percentage of the oxygen to change from breathable O_2 to the extremely poisonous O_3, or ozone. Ozone's absorption of ultraviolet sunlight heats the atmosphere and counteracts the falling temperature. Hence a minimum temperature occurs in the stratosphere some 17 kilometers up, and a maximum at some 50 kilometers where nearly ground temperatures are recovered. Another very cold minimum occurs near 80 kilometers, above which the temperature begins to rise again. Above 300 kilometers a trace of atmosphere is left at a temperature of some 1500°C, heated by solar ultraviolet light and high-energy particles.

Above some 1000 kilometers we encounter the famous Van

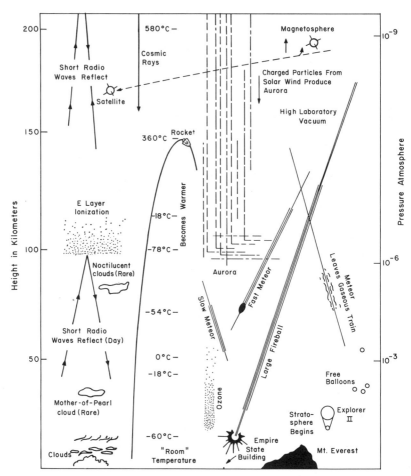

Fig. 59. Phenomena in the upper atmosphere. (Drawing by Joseph F. Singarella.)

Allen belts (Figs. 60 and 61), named for their discoverer, James A. Van Allen, who detected them by measures from early U.S. satellites. They may be classed as parts of the Earth's atmosphere, consisting of very energetic ionized nuclei of atoms, mostly hydrogen, and electrons trapped by the Earth's magnetic field. The Van Allen belts are very dangerous to living organisms not protected by shielding equivalent to about a half inch of lead, which some of the particles can penetrate. Some of the Van Allen particles are produced by cosmic-ray collisions in the atmosphere and the major portion by violent sprays of particles from the Sun in solar flares. They can also be introduced by nuclear explosions at great heights. From extreme parts of the belts the particles leak out fairly rapidly, in days or hours, while in the central part individual particles may persist for much longer times.

The air density decreases with height, reaching about one-millionth the surface value near a height of 95 kilometers. Half of the air is contained in the first 5.6 kilometers above the surface, half of the remainder above 11.3 kilometers, and so on. At the elevations where meteors are seen, radio signals are reflected, and the aurora borealis appears, the density of air is no greater than in the vacuum of a thermos bottle. A much thinner atmosphere than ours would still be a good shield from external dangers, but might not regulate the surface temperature so efficiently.

The interior of the Earth cannot, of course, be studied as easily as the atmosphere but there are methods of learning a great

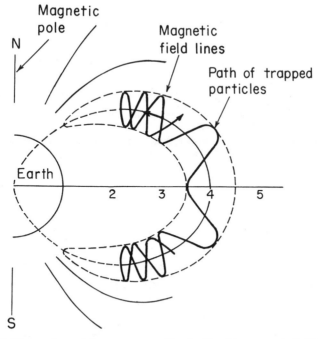

Fig. 60. Charged particles are trapped in the Earth's magnetic field because, as shown above, they are forced to turn as they move across the field lines. Numbers on the horizontal line indicate distances in Earth radii.

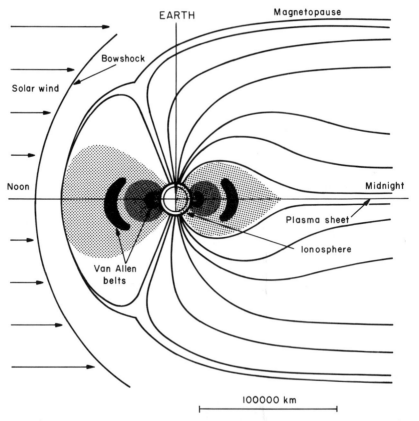

Fig. 61. The Earth's *magnetosphere* is maintained by the solar wind, from the left in the diagram. The interaction of the charged ions and electrons in the magnetic and electric fields produces a complex and variable plasma structure.

deal about it. Deep wells have penetrated only about 10 kilometers, a negligible fraction of the distance to the center. For vital information about the gross distribution of mass in the Earth we turn to artificial satellites and to measurements of the effects of the equatorial bulge on the motion of the Moon. Geological data are highly important in describing the surface layers. Variations in the attraction of gravity from place to place and the strengths and directions of Earth magnetism add valuable data concerning somewhat lower levels, but the most exact information about the layers deeper than a few kilometers proceeds from the manner in which earthquake (or seismic) waves travel through the Earth. To describe even briefly the many methods used to study the Earth would require no less than another book. A short outline of the results, however, will impart some impression of the Earth's construction and will demonstrate the precarious nature of the existence we enjoy on the skin of this planet.

Astronomical results give the average density (5.52 times the density of water) of the Earth and its shape at the surface; the attraction of the equatorial bulge and the known densities of rocks furnish a consistent idea of the densities of the Earth near the surface—about 2.6 times the density of water, or half the mean density.

The deep wells show that the temperature generally increases with depth at an average rate of about 1 degree centigrade in 50 meters, although the rate varies tremendously from place to place. If this temperature increase were to continue to the center of the Earth, the temperature would reach the very high value of 130,000°C. That this temperature must be far too great is shown by geophysical investigations, which place the central temperatures at about 6000°C. The low heat conductivity of the surface rocks makes the change of temperature with depth very rapid near the surface, while in the deep layers heat is better conducted and the temperature changes more slowly. Radioactive elements such as uranium and thorium, which seem to be concentrated largely in the Earth's crust, add to the heating of the outer levels. The residues of the radioactivity, among which are helium and certain leads, provide a measure of the age of the rocks in which they are found and of Earth itself—some 4.6 thousand million years, or 4.6 *aeons*.

Geophysical evidence indicates that the Earth was once heated so that the interior was melted. As the short-lived radioactive atoms disintegrated, the entire planet cooled and most of it solidified. In the earlier stages, more than 4 aeons ago, the rate of heat loss must have been very much greater than at present because swift convection currents in the interior would have carried the heat up rapidly. After the formation of a solid crust the rate of heat loss became relatively slow, volcanoes and lava flows carrying at most a few percent of the total heat.

Without the information gained from the study of earthquake waves the structure of the deep interior of the Earth

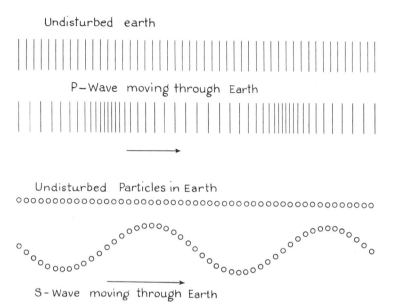

Fig. 62. Earthquake waves. The *P*-waves move by condensation and rarefaction, while the *S*-waves are transverse vibrations.

would be mostly conjecture. When an earthquake is produced, usually by a sudden slipping of the crustal rocks along a fault sometimes hundreds of kilometers below the surface, two main types of wave are sent out in all directions through the Earth. The *P*-wave, *primary* or *pressure* (or condensation) wave, is like a sound wave. The vibration is carried forward by a condensation of higher pressure and density, as in Fig. 62 (*upper*). The second type of vibration is the *S*-wave, *secondary* or *shear* wave, in which the motion is perpendicular to the direction of travel, as for a light wave or a wave on the surface of water (Fig. 62, *lower*). The *P*-wave, always traveling faster (approximately 1.8 times) than the *S*-wave, makes an earlier record of an earthquake on the seismograph at the recording station (Fig. 63). A few kilometers below the Earth's surface the *P*-wave travels at about 8 kilometers per second and the *S*-wave 4.6 kilometers per second. The speeds of both waves increase at greater depths, where the densities and pressures are greater. The destructive energy of an earthquake is carried by slower surface waves (*L*-waves), more complex in nature than the *P*- and *S*-waves.

The most remarkable result obtained from a study of the rec-

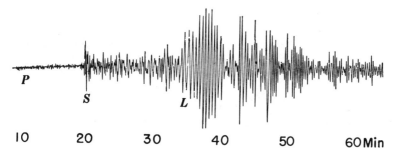

P

S   L

10   20   30   40   50   60 Min

Fig. 63. A seismogram, showing the nature of the record made by the various earthquake waves. The *P*- and *S*-waves precede the stronger surface *L*-waves. (Courtesy of L. Don Leet.)

ords made by these waves after they pass through various parts of the Earth is that the *S*-waves do not penetrate a central core (the Dahm core), which extends slightly more than half way out from the center of the Earth (Fig. 64). Since *S*-waves are damped out in liquids, the central core must be liquid having a density of about twice the average for the Earth. This liquid core contains an inner solid core of even higher density, extending some 1300 kilometers from the center.

The pressures near the center of the Earth are tremendous, about 3.7 million atmospheres. It is difficult, therefore, to predict precisely at what temperature any given material will melt or how much it will be compressed. The compression of a solid or liquid is certainly considerable near the center and the melting temperature will certainly be high. Since the Earth is magnetic and since iron and nickel-iron are so prevalent in meteorites, which are our only samples of material from other planetary bodies, most investigators believe that the core of the Earth is made largely of iron or nickel-iron mixed with lighter elements such as sulfur and silicon or oxygen. The high pressure compresses the iron from a density of 7.7 that of water, which it has on the Earth's surface, to about 10 to 12 at the center. There the *P*-waves travel about 11 kilometers per second.

Just outside the inner cores in the intermediate shell as seen in Fig. 64, the densities are about the average for the entire Earth. The pressure compresses ordinary rocky material to this density. The outer mantle is somewhat denser (4.3 times water) than the heavier rocks, but may be largely composed of them. The crust consists mostly of granites and other igneous rocks,

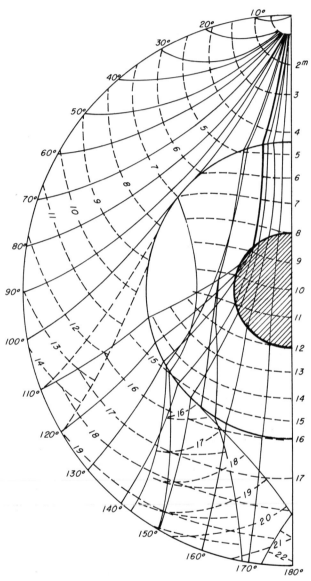

Fig. 64. Paths of earthquake waves in the Earth. The changing density with depth causes the paths to curve. At the discontinuities of the core boundaries the plotted *P*-waves are bent abruptly. The *S*-waves (not plotted) do not penetrate the core and are shadowed from stations opposite the earthquake. The numbers represent minutes of travel time. Note the solid inner core.

the sedimentary rocks appearing only in the upper kilometer or so, on the average.

The outer crust of the Earth varies in thickness from place to place and from investigator to investigator, but appears to be some 40 kilometers thick under the continents and as thin as 6 kilometers at some spots under the oceans. Geologically the crust appears to float on a deformable but exceedingly viscous layer, the asthenosphere, 1 to 6 hundred kilometers thick. This is part of the Earth's *mantle,* which extends from the bottom of the crust to the top of the liquid core. The formation of mountains and the extensive distortions of the crust show clearly that the crust is subject to motions that could not be possible unless there were an underlying layer of pseudo-liquid material that can yield. According to the principle of *isostasy* the total mass under any given area is constant. Lighter surface materials, such as those that form mountains, are lifted. Glaciers depress the surface, which again rises slowly after the ice melts. For quick-acting forces, such as earthquake waves, the material is very rigid; to forces acting over long periods of time it gives way. Glass is such a material. "Silly Putty," a silicon compound that bounces on impact but kneads like putty and flows like water in a few hours after deformation, simulates the properties of the solid earth, on a very short time scale. Volcanoes show, of course, that some liquid material must exist just below the crust, but all of the deformable layer need not be liquid in the ordinary sense.

In recent years extensive mapping, sampling, coring, and temperature measuring of the oceans bottoms have been made, particularly by the Woods Hole Oceanographic Institute in Massachusetts and the Scripps Institute at La Jolla, California. Down the whole length of the Atlantic Ocean occurs a high ridge with a deep rift near its peak (Fig. 65). Such a system of ridges, some 2000 meters high, and usually with a rift, occurs in most oceans. Generally the temperature gradient in the ocean bottom is greater near the rifts, indicating a greater heat loss there and a thinner crust. Horizontal faults, tens to hundreds of kilometers in length, often break the continuity of the rifts. Recently submarines have been able to study the deep ocean bottoms along the rifts and in some have discovered startlingly abundant and new species of sea life. The heat from the under-

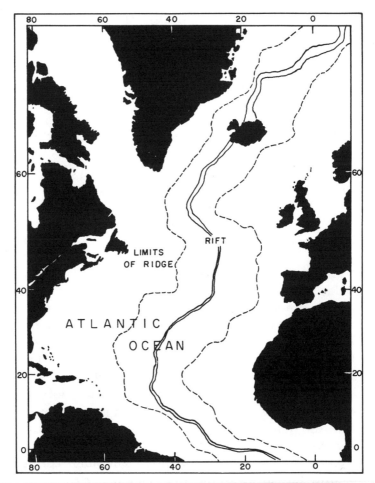

Fig. 65. The Atlantic ridge, with the floor of the rift valley 3,000 to 5,000 meters below sea level flanked by ridge peaks submerged only 1,000 to 2500 meters. Deep earthquakes occur systematically along the rift. The numbers are degrees of latitude and longitude. (After Bruce C. Heezen.)

sea lava warms the water locally while frozen lava is profuse along the rifts.

Two pieces of geological evidence—climatic changes over the last 500 million years and records of the ancient magnetic field in old sediments and lavas—indicate that the crust of the Earth has slipped or slid around the interior like the shell of an egg. This motion is reminiscent of the classical method of distiguishing between fresh and hard-boiled eggs without breaking the shells: a spinning fresh egg that is stopped and suddenly released will start spinning again. The North Pole may once have been in the Pacific Ocean. The Earth's pole does have a wobbling motion of a few feet as measured by changes in the latitudes at various stations and possibly may be moving systematically (Fig. 66).

That the continents have moved with respect to each other no longer remains uncertain. Figure 67 illustrates the manner in

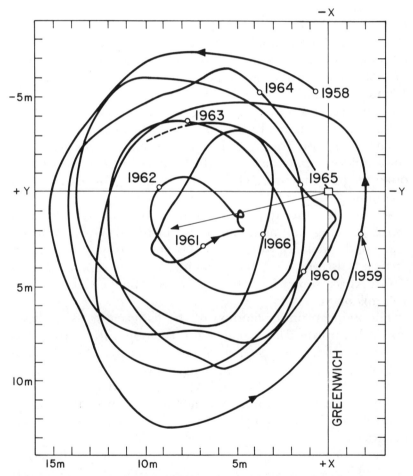

Fig. 66. Motion of the North Pole over the Earth's surface from 1958 to 1966. The rectangle measures the mean starting point in 1903, and the arrow indicates the average direction of motion, some 10 centimeters per year. (Courtesy of the Smithsonian Astrophysical Observatory.)

which the various continents appear to have developed from a single continent, as suggested by Alfred Wegener in 1924. The fit of the "jigsaw-puzzle" pieces is supported by new evidence in addition to biological evidence. The oceanographic studies, mentioned above, show that sediments are thinner and newer near the mid-oceanic ridges, where earthquakes abound. Certain old rock formations in Africa and South America, as well as in Greenland and the British Isles, match up well with the boundary shapes shown in Fig. 67. The ocean bottoms near the ridges are spreading apart—very slowly, a few centimeters per year. It was the proof of this spreading by F. J. Vine and Tuzo Wilson in 1966 that finally convinced geoscientists of the reality of continental drift. Vine and Wilson found that the magnetic polarity in strips normal to the oceanic ridge near Vancouver Island were matched in both directions outward from the ridge. As new magmas welled up along the ridge, they froze in the variable magnetic field of the Earth. With them the magnetic polarity was also frozen. All of the oceanic ridges measured show this same phenomenon, which can be cross-dated with the other magnetic records. The continents have broken away from a common supercontinent, called Pangaea, some 200 million years ago. It rather quickly broke up, in only about 20 million years, forming Laurasia, which also included North America, and Gondwanaland. By this time the latter was splitting off Antarctica, Australia, and India. South America separated from South Africa somewhat late so that the Atlantic Ocean is only 160 million years old!

The modern understanding of the Pangaea breakup, based on vastly improved rock dating information, differs from Wegener's original concept (Fig. 67) chiefly in that the first large rift occurred along the curve lying between North and South America and between Asia and Africa rather than between the Americas and Eurasia plus Africa. India apparently "shot up" from Australia, "banging" into Asia, and so produced the Himalayas.

The basic cause of the oceanic rifts, continental drift, mountain making, volcanos, and other large crustal activities on Earth is now well understood although many details are obscure. The subject is known as *plate tectonics,* because very large crustal plates, of continental dimensions, are moved by heated circulating currents under the crust. An upwelling current spreads out

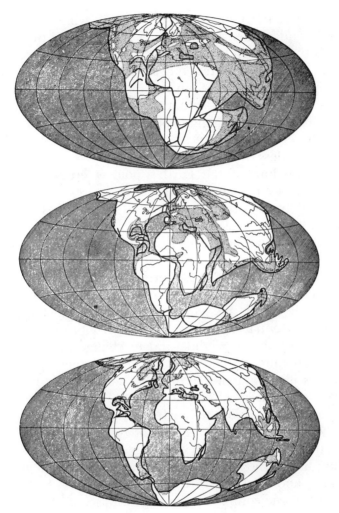

Fig. 67. Continental motions according to Wegener's theory. (From *Lehr-buch der Physik;* courtesy of Friedrich Vieweg und Sohn.)

or peaks as it forms an oceanic ridge, carrying the newer frozen crust with it perpendicular to the ridge. When an oceanic plate is forced toward a continental edge, the denser ocean bottom crust plunges below the lighter continental crust, and crumples the edge of the continent. An island arc such as the Japanese islands or Aleutians may result. When two continental plates are pushed together, great mountains result—the Himalayas are an example. So long as the dying radioactivity keeps the Earth's interior hot enough the continents will drift and change, volcanos will be born and become dormant, while mountains will be built and eroded away. On an astronomical time scale the Earth's surface is in turmoil. We are beginning to see it in slow motion, however. Lasers acting as radars on reflecting satellites, space probes, and the Moon will measure intercontinental distances to accuracies of centimeters and show us detailed motions sooner than we might expect. Observations of artificial satellites have already reduced these errors in distance from a hundred to less than one meter.

The wobbling of the Earth's pole where it "pierces" the Earth's surface is a small effect, but of interest because it is well explained by theory. The principal motion has a period of about 430 days, while there is a smaller motion in a year (see Fig. 66). Seasonal changes cause a melting and shifting of the ice in the polar caps which would produce a small yearly effect, but the 430-day period requires more explanation. If we again consider the Earth as a top, it is like one that was set spinning badly, not exactly about the symmetrical axis perpendicular to the plane of the equatorial bulge. Major earthquakes may also tend to displace the pole slightly and to change the nature of the polar wobbling. If the Earth were absolutely rigid, the pole would oscillate in about 10 months, but since the Earth is only twice as rigid as steel, the period is 430 days, as observed.

The fragile crust of the Earth, floating on the heated and deformable rocks underneath, is not the stable and permanent layer that it appears from everyday experience. Not only is it drifing about the main body of the Earth, but it is certainly cracking and buckling through geologic ages. Many regions, having become completely covered with ice in the glacial ages, sank under the load. When the ice melted, they rose again. The magnetic field keeps wandering around and occasionally changing sign while the continents are shifting and changing shape. The crust is in continuous vibration produced by earthquakes. Volcanoes may become active at any time and occasionally produce catastrophic results, such as the violent explosion of Krakatau. Furthermore, great meteorites occasionally fall and can devastate huge areas of the Earth.

When we contemplate all of these dangers to life both from within and without the Earth, we must indeed marvel that we still exist—but, of course, if we did not . . .

7

The Moon's Influence
on the Earth

The era is well past when mystical powers of the Moon were supposed to influence our everyday life on the Earth. No longer do thinking people attempt to credit the Moon with their successes or blame it for their failures. The Moon does, however, influence the Earth directly in many ways—all subject to simple laws of physics and dynamics.

The Moon is so large and so close to us that it reflects sufficient sunlight at its full phase to light up the night satisfactorily for many practical purposes of life. It is massive enough to distort the shape of the Earth and to produce tides in lakes and oceans. It provides the main force that moves the poles of the Earth in the precession of the equinoxes. Its distortion of the Earth's shape produces friction that slowly lengthens the hours of the day. Its shadow on the Earth at occasional places and times obscures the light of the Sun to produce solar eclipses. In such ways our nearest neighbor in space makes its presence known. To see how these effects are brought about, let us begin

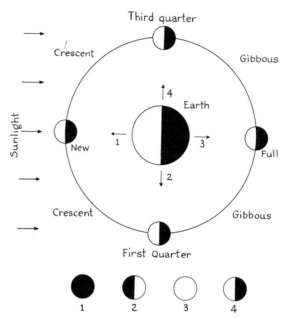

Fig. 68. Phases of the Moon as seen from the Earth during one synodic month.

by investigating the motions and superficial appearance of the Moon.

The calendar month approximately represents the Moon's period of revolution about the Earth. Were fractional months feasible in a calendar, there should be 12.37 . . . months per year, because their average length is 29 days 12 hours 44 minutes 2.8 seconds. This period, technically the *synodic* month, is the space of time in which the Moon passes through its sequence of phases from *new* to *first quarter,* to *full,* to *third quarter,* to *new* again (Fig. 68), and makes a complete revolution about the Earth with respect to the Sun. Since the Earth has moved forward about 30° in its orbit during this time, the true or *sidereal* month, measured with respect to the stars, is a little more than 2 days shorter than the synodic month. On the average, a sidereal month has a length of 27 days 7 hours 43 minutes 11.5 seconds.

The reason for the difference in lengths of the two months can be seen from Fig. 69. Starting from a new moon, *A*, when Sun, Moon, and Earth are in line, we see that the Moon returns

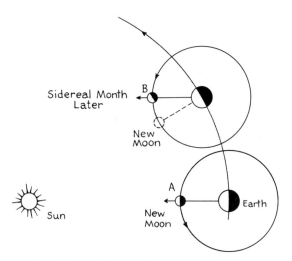

Fig. 69. A sidereal month is shorter than a synodic month. The Moon is not yet new again at position B, although it has completed a revolution about the Earth with respect to the stars.

to the same direction with respect to stars *before* it reaches the *new* phase again, because the Earth has completed part of its revolution about the Sun in the meantime.

The most curious fact about the Moon's motion is that the Moon rotates on its axis at the same average rate that it revolves about the Earth. Thus we always see the *same* hemisphere of the Moon's surface and *never* can see the other hemisphere. To demonstrate this motion, hold a ball or globe rigidly at arm's length and slowly turn around. As your body makes one revolution so does the ball, but you see only one side unless you turn the ball in your hands.

Because the Moon is the nearest celestial object, its distance is the most accurately known, to a few centimeters by laser or lidar. The nearest that the Moon can approach the Earth's center is some 356,410 kilometers. An observer, being located on the surface of the Earth, may move some half Earth diameter closer than this (Fig. 70). The greatest distance that the Moon can attain is 406,700 kilometers, while its mean distance is 384,401 kilometers.

The Earth's atmosphere has a surprising effect upon observations of the rising or setting Moon. Light rays are bent by the atmosphere to such an extent that the *entire Moon (or Sun) can be*

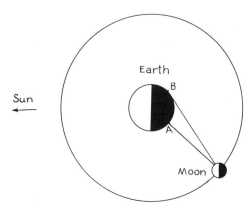

Fig. 70. The Moon's distance is less for observer *A*, who sees the Moon overhead than for *B*, who sees it setting.

seen before it has risen and after it has set. The *refraction* of the light coming from empty space into the atmosphere is just about 0.5 degree, the apparent diameter of the Moon. Thus, when the Moon's upper limb would be just out of sight were there no atmosphere, the entire Moon is apparently lifted into view (Fig. 71). At greater heights the refraction is less, and it decreases to zero overhead.

Everyone has noticed the strange phenomenon that the Moon appears to be larger when seen near the horizon than when seen overhead. Actually, when *measured*, the diameter is smaller when near the horizon, because of the small distance effect mentioned above and because refraction flattens the disk slightly. The standard explanation has been that the Moon

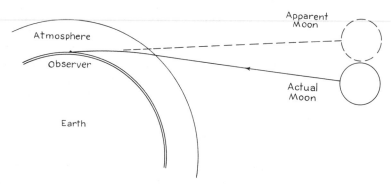

Fig. 71. Refraction in the Earth's atmosphere apparently raises the Moon (or Sun) above the horizon after it has set geometrically.

Fig. 72. The Moon illusion: the two black disks are of equal diameter. (After a diagram by I. Rock and L. Kaufman.)

seems larger when seen in juxtaposition to distant objects on the horizon than when seen against the expanse of the sky. Since the effect is the same for an unbroken horizon at sea as for a land horizon, this explanation is not satisfactory. Psychologists show that the effect arises from a peculiar property of the brain and eye. The observer tends to visualize the Moon as more distant when near the horizon than when overhead. That is, he unconsciously places it at the distant horizon. Since the Moon remains unchanged in angular size it *seems* to be intrinsically larger near the horizon where it *seems* to be more distant (Fig. 72).

Important as the Moon may be as an object for stimulating the human mind, its greatest effect on the Earth results from its power to produce tides. This power is a direct consequence of the gravitational attraction of the nearby Moon for the Earth.

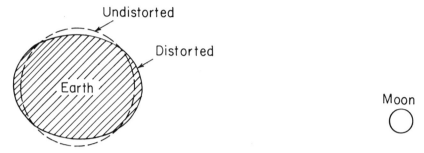

Fig. 73. The cause of tides. The Moon elongates the Earth along their line of centers.

The cause of the tides was early appreciated by Newton as a confirmation of his law of gravity. Since the attraction is inversely proportional to the square of the distance, the part of the Earth nearest to the Moon is attracted by a force nearly 7 percent greater than the part farthest away. The force at the center is, of course, the average value, which is exactly sufficient to hold the Moon in its orbit. The 7-percent differential in force acts on the body of the Earth as a distortion, tending to stretch the globe along the line joining it with the Moon (Fig. 73).

A most interesting feature of this tide-raising force is that the face of the Earth *away* from the Moon is distorted in almost exactly the same fashion as the face toward the Moon. One may understand this symmetric elongation by considering that the lunar hemisphere of the Earth is pulled away from the center, and that the center is pulled away from the opposite hemisphere. When the Earth is stretched along the line joining it to the Moon, the circumference perpendicular to this line is naturally compressed. The net tendency of the tide-raising force is to distort the Earth into a shape similar to that of a symmetric egg.

Now if the Earth were absolutely rigid, not yielding to the distorting forces that act on it, all of the tidal effects would occur in the oceans and surface waters. If the Earth were perfectly elastic with no rigidity, the ocean tides would be negligible, although the tidal bulge would still exist. The comparison of ocean tides with the predicted values is exceedingly difficult, however, because the measured tides at shore stations depend upon currents that are set up over the irregular ocean beds. Careful measures of the tides in long pipes show that only 70 percent of the theoretical effects actually occur. The main body of the Earth yields to the forces to the extent of the remaining 30 percent. From these measures it is deduced that the Earth as a whole is more rigid than steel. We have seen that the data from observations of earthquakes and from the motion of the Earth's poles confirm this result. Outside the liquid core the Earth has an average rigidity about twice that of steel.

A surprising additional result about the Earth was found from the tide experiments. *The Earth is an elastic ball.* Before the experiments, it was generally believed that the Earth was viscous, like thick molasses or glass; if it were distorted a small amount it would probably remain so or else *slowly* regain its original shape because of the small restoring forces. The experiments showed that the entire Earth yields *immediately* to the tide-raising forces, in so far as its rigidity will allow, and that it *immediately* returns to its original shape when they are removed. Thus the Earth is not only more rigid than steel; it is also more elastic.

Although the Moon is the most powerful body in raising tides on the Earth, the Sun also is an important contributor—to the extent of about 30 percent. The Sun produces tides in exactly the same manner as the Moon. When the two bodies are nearly in line, as at new or full moon, their tidal forces add together. When their directions are at right angles, as at first or third quarter, their tidal effects tend to cancel. The result is that at new or full moon there are *spring* tides, in which the high tide is very high and the low tide is very low. In between, at first or third quarter, there are *neap* tides, in which the range from high to low tide is reduced to less than half the value at spring tides.

Still another factor enters into the production of the tides. When the Moon is nearest to the Earth, at *perigee,* its tide-raising force is greater than when it is farthest away, at *apogee.* The range in the lunar part of the tide changes by about 30 percent because of this change in distance.

Although the prediction of the theoretical tides is somewhat complicated, the prediction of actual tides at a given station is even more difficult. The tides at shore stations generally have ranges of meters, considerably above the average expectation by simple theory. This discrepancy arises from the fact that the

Fig. 74. Unequal daily tides. The tide at *A* will exceed that at *B*.

observed tides are measured at the shallow edges of the oceans. As the Earth rotates, the tidal bulges of Fig. 73 become, in effect, tidal waves, which pile up on the sloping ocean beds near the shores, much as ocean swells may grow to high waves as they approach a gently sloping beach. In the Bay of Fundy, where this effect is further augmented by the occurrence of a funnel-shaped shore line, the tidal range is often 16 meters or more. In many places the high tide is consistently later than the maximum tide-raising force by several hours. This observed delay is known as the *establishment* of the port, and is used in predicting the tides.

The obliquity of the ecliptic has a marked effect on the tides at stations away from the equator. Because the Earth's poles are tipped from the plane of the Moon's revolution, the two daily tides may differ greatly in range. By reference to Fig. 74, one can see that the tide at point *A* will be greater in range than the one at *B*, half a day later. At certain stations it often happens that only one high tide instead of two will be appreciable.

Practically no living person can fail to be impressed by a second remarkable phenomenon caused by the Moon—a total eclipse of the Sun. At rare intervals, because of a striking coincidence, the Moon is at just the proper position to blot out the Sun's light for a small area of the Earth. If the Moon were a bit nearer, solar eclipses would be commonplace, and if it were a bit farther away we could never see a total eclipse.

In Fig. 75, the Moon's shadow on the Earth is shown. For an observer inside the dark cone (the *umbra*), no direct rays of the Sun are visible; only the *corona*, the outermost vaporous atmo-

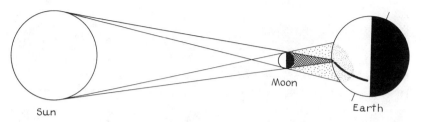

Fig. 75. Total solar eclipse. The dark shadow is the umbra of total eclipse and the shaded area is the penumbra of partial eclipse. A path of totality is shown. The relative dimensions are much exaggerated.

sphere of the Sun, and high prominences, can be seen. Outside the umbra, in the partially darkened shadow (the *penumbra*), a part of the Sun's disk is covered. As the Moon's shadow passes over the Earth's surface, the sunlight is slowly dimmed during a period of an hour or more (Fig. 76). As the light becomes weaker, with only a thin crescent of the Sun's disk exposed, a phenomenal coolness and silence prevails. The crescent is still so brilliant that it must be viewed through a dark filter. Just before totality the narrowing crescent breaks up into a series of beads, as the last rays from the Sun shine through the valleys of the precipitous lunar surface. These *Baily's beads* show brilliantly for a few seconds only (Fig. 77). By this time a glowing ring can be seen completely around the Moon, and if, as occasionally happens, one bead alone is bright, the effect is that of a luminescent diamond ring. Oddly enough, these beads really should be called Williams' beads, since Samuel Williams (1743–1817) observed and described them many years before Francis Baily (1774–1844) did.

A total eclipse can never continue for much over 7 minutes and usually lasts for a considerably shorter time, yet it is a sight that well repays an observer for his effort in traveling to the zone of totality. In any given location a *total* eclipse can be seen, on the average, only once in 360 years. Sometimes, only the penumbra of the Moon's shadow strikes the Earth, and produces a partial eclipse, while in other eclipses the Moon is so far away that its disk does not completely cover the Sun. In the latter case the eclipse is *annular* or ringlike.

Eclipses of the Moon by the Earth's shadow are less numerous than solar eclipses but each is observable over more than half of

Fig. 76. Solar eclipse of February 1961. (Photographed by A. H. Catá of Geneva, Switzerland; courtesy of *Sky and Telescope.*)

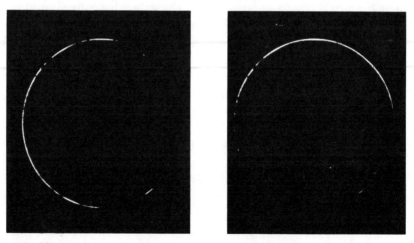

Fig. 77. Baily's beads. (Photograph by the Yerkes Observatory.)

the Earth's surface. Consequently, at a given position on the Earth, a lunar eclipse can be seen quite frequently. In some years none occur, while the maximum number is three. The maximum number of eclipses in one calendar year is seven, five solar and two lunar or four solar and three lunar. A lunar eclipse, however, is not at all spectacular, and it has little value to the astronomer. When the Moon lies completely in the umbra of the Earth's shadow, it usually acquires a dull copper hue because some sunlight is refracted through the Earth's atmosphere to produce a sunset effect. On rare occasions, the eclipsed Moon is very dark, because the Earth's atmosphere is clouded completely around the twilight zone; at other times part of the Moon is illuminated during totality.

During a solar eclipse the astronomer has an opportunity to observe the corona, which is an extended but exceedingly rarefied and hot mantle of gas about the main body of the Sun (Fig. 78). Also, the *prominences* of the Sun, great whirling or exploding clouds of hydrogen and calcium gases, can be seen (Fig. 79), though they and the corona can now be well observed without an eclipse. The astronomer has also an opportunity to photograph and measure the positions of stars near the Sun, where no measures can ordinarily be made because of the brilliancy of the sunlight scattered in the Earth's atmosphere. These measures have shown that the light from distant stars has been minutely deviated by the Sun's mass, in accordance with the predictions of Einstein's theory of relativity. This demonstration, coupled with the anomalous motion of Mercury's perihelion, constitute two of the three astronomical verifications of the relativity theory.

The ancient observations of solar eclipses have been invaluable in showing that the Moon is tending to increase the length of the day by acting as a brake on the Earth's rate of rotation. The lack of accurate time-keeping devices in antiquity has been no handicap in this type of investigation, because the *place* where a total eclipse of the Sun could be observed is, in itself, a good measure of the time and of the position of the Moon when the eclipse occurred. The Earth must be turned at a certain angle and the Moon must be in a specified position for the Moon's shadow to fall on a given point of the Earth's surface. Calculations based on the records of ancient eclipses show that

Fig. 78. The solar corona at the total solar eclipse of August 31, 1932, Maine. (Photograph by the Lick Observatory.)

the day is *increasing in length* by nearly 0.001 second every century. This change arises from tidal friction.

The energy of the tides is changed to heat, at a rate of some 4 billion horsepower. In 1920 Sir Harold Jeffreys calculated that the friction of the moving water in the shallow areas of the Bering and Irish Seas accounted for 80 percent of the observed change in the length of the day. Present-day theories distribute the tidal friction much more evenly throughout the oceans.

Some remarkable research by John W. Wells shows that tidal

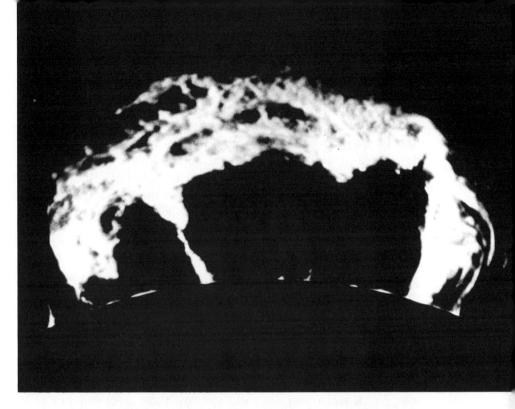

Fig. 79. Solar prominences are giant clouds of incandescent hydrogen, calcium, and other gases. They take many forms, which are controlled by magnetic fields on the Sun. (Photographs by the Sacramento Peak Observatory,
Air Force Cambridge Research Laboratories.)

friction is not limited to the historical period of man but has been in progress since ancient geological periods as far back as the Devonian, nearly 400 million years ago. He finds that certain fossil corals of this period exhibit *annulations,* like tree rings, that measure daily and annual growth rates. Thus the year then contained some 400 days, reducing the day to some 22 hours. C. T. Scrutton has since found *monthly* annulations, indicating that the lunar month was then only 21 days in length. We shall return to these extremely important facts in discussing the evolution of the Earth and Moon.

The accurate observations of the Moon, Mercury, Venus, and the Sun during the past century demonstrated strikingly irregular variations in the length of the day. In Fig. 80 the deviations in the observed positions of these bodies are reduced to the equivalent time error in the position of the Moon. Since the deviation curves are nearly identical, it must be concluded that the Earth is a bad clock. In the latter part of the nineteenth century, the Earth ran relatively fast by more than a second per year. After 1900 it ran slow by less than a second per year.

A rate of 1 second per year does not particularly exceed the accuracy of the best pendulum clocks and is far below the accu-

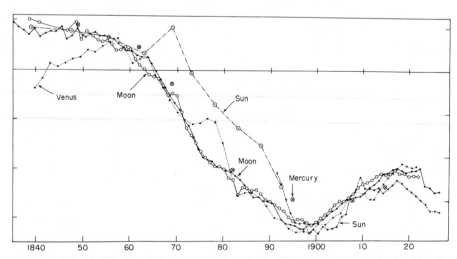

Fig. 80. The Earth is not a perfect clock. The curves above, derived by H. Spencer Jones, represent observed deviations from the calculated motions of the Moon, Sun, Venus, and Mercury. Since these bodies could scarcely deviate by chance in the same fashion, the times of observation must be in error. Therefore the Earth turns irregularly.

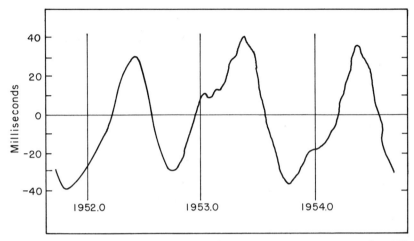

Fig. 81. Seasonal changes in the time of the Earth's daily rotation (After William Markowitz, U.S. Naval Observatory.)

racy of modern atomic clocks. It is a tremendous rate for the Earth considered as a rigid body. On the other hand, if the Earth's radius were to expand or contract uniformly by only centimeters, the observed errors could be explained. There is some evidence that the changes in rate are associated with the occurrence of deep earthquakes. Such an association is not surprising in consideration of the fact that some type of alteration must take place in the Earth to cause its rate of rotation to vary.

In recent years the vibrations of quartz crystals, of atoms in ammonia molecules, of caesium atoms, and of other atoms have been counted and harnessed to operate clocks. The precision attained with such clocks is remarkable and increases continuously with technological advances. Errors of 0.001 second per year are considered large, while improvements to the extent of another factor of 10,000 times are underway. Compared to these clocks the Earth keeps time like a cheap watch. Seasonally on the Earth, the systematic winds, meteorological phenomena, solar tidal changes in the crust, and other minor effects vary the Earth's rate of rotation, producing clock errors like those shown in Fig. 81.

Even though the Moon does not affect our lives by direct subtle forces, it certainly adds variety to our lives by way of tides and eclipses. And at times viewing the Moon can be a truly inspirational experience.

8

Observing the Moon

As we begin this survey of the Moon's surface, we meet the problem that will confront us repeatedly in planetary studies made from the Earth—the problem of observing fine detail by means of a telescope. The telescope is the key of astronomy, an instrument of great precision with which astronomers unlock the closed doors of the universe. But even a perfect telescope when used on Earth by a most expert observer is limited by an insuperable handicap, the Earth's atmosphere. This is one reason that the space program has been so popular with astronomers.

Our knowledge of the surface features of the Moon (Fig. 82) or of the planets has, until recently, been derived only from a study of reflected sunlight, which, before orbiting observatories became possible, could reach us only after passing the ocean of atmosphere above. We have seen that refraction in this atmosphere bends the light rays through a small angle; unfortunately, no two parts of the atmosphere refract exactly the same

Fig. 82. The Moon past first quarter. The north pole is at the top. (Photograph by the Lick Observatory.)

amount. As winds and currents of warm and cool air circulate overhead, each ray of light is bent in a slightly different manner. The result is apparent to the naked eye: the stars twinkle. The planets, however, rarely twinkle because they have finite disks; hence they can often be recognized by their steadiness. A telescope magnifies the twinkling so much that often stellar images appear to "boil," as though they were seen across the surface of a hot stove, or across heated desert sands. This boiling turbulence of the atmosphere produces "seeing," which may be relatively good or bad, the quality depending upon the appearance of stellar images in a telescope.

Besides distorting the images of stars, the atmosphere steals away some 30 percent of the incoming light and scatters it in all directions. Above the atmosphere the sky is much blacker at night and equally black in the daytime; thus the stars and planets can be observed by day as well as during the night. On a clear day when Venus is near its brightest it can be seen by the naked eye—if one stands in a shadow and knows exactly where to look. The brightest stars have been reported as visible in the daytime if observed through a tall chimney, a mine shaft, or the like. The reader, however, is advised to use a small telescope or binoculars to see stars in the daytime.

Only a few people have had the opportunity of seeing a black sky during a clear day. The first was Major Albert W. Stevens, who from the stratospheric balloon Explorer II in 1935 observed the appearance of the sky as seen at an altitude of 22 kilometers above sea level: "The horizon itself was a band of white haze. Above it the sky was light blue, and perhaps 20 or 30 degrees from the horizon it was of the blue color we are accustomed to. But at the highest angle that we could see it, the sky became very dark. I would not say that it was completely black; it was rather a black with the merest suspicion of dark blue . . . To look directly at the sun through one of the portholes was blinding. The sun's rays were unbelievably intense." The astronauts, completely beyond the atmosphere, now report similar observations from the vacuum of space.

The brightness of the night sky as seen from the Earth affects visual observations only slightly, because the eye is not sufficiently sensitive to be blinded by such weak light. The photographic plate or image tube, however, can be exposed until it is overwhelmed by the night-sky light. Our atmosphere, therefore, handicaps tremendously the photography of faint nebulous objects, which may be much fainter than the diffuse sky light.

The bad seeing sets an insurmountable barrier to observing fine detail, either visually or photographically, on the Moon and planets. Below a certain angular limit neither the eye nor the photographic plate can register any detail. This limit is at best about 0.1 second of arc and corresponds to the theoretical *resolving power* of a telescope with an aperture of 1.1 meters. The theoretical resolving power varies inversely as the aperture of the telescope and hence becomes 0.5 second of arc for a 23 centimeter aperture. For bright objects the eye is more effective than the photographic emulsion because it can register details during those rare instants when the seeing is nearly perfect. The photographic plate, on the other hand, requires an appreciable exposure time, during which the seeing will change. The resolving power of lunar photographs rarely exceeds 1 second of arc, or about 1.6 kilometers. Thus a relatively small telescope, under good seeing conditions, can reward the patient observer with an extremely good view of lunar details. Television image tubes, having much greater sensitivity than the photographic emulsion, can sometimes match the eye in detecting fine details.

To minimize the undesirable atmospheric effects, astronomers have searched to "the ends of the Earth," and now beyond, to find locations where the seeing is exceptionally good. Mountain tops, above the dust and water vapor of lower areas, generally provide very transparent skies, but the seeing on a mountain chosen at random may be poorer than at sea level. The best observing sites in the world now appear to be in the Chilean Andes. High-altitude balloons and aircraft are extremely helpful but telescopes in space offer the only ideal solution to the problem of seeing.

As we have noted, a large telescope will enable an observer to discern finer details than a small telescope—if the seeing permits. When the seeing is bad, however, the image of a planet or the Moon in a large telescope may be even poorer than in a small one, because the larger area allows a greater variation of the air conditions. Hence telescopes of moderate aperture (15 to 50 centimeters) are usually the most effective for direct visual studies. The great reflectors are used almost exclusively with registering devices where their tremendous light-gathering power is of the utmost value. The magnifying power for any telescope is the ratio of the focal length of the objective to that of the eyepiece and may be chosen at will by a change of eyepieces. A high power is used when the seeing is good and a lower power when it is poor.

The Moon is a spectacular object as seen through any telescope, whether large or small. Galileo was the first man in his-

Fig. 83. Rays at full moon. Note the magnificent rays from Tycho (*lower center*) and from Copernicus and Kepler (*middle left*). Note also the many craters that show bright rims. (Photograph by the Lick Observatory.)

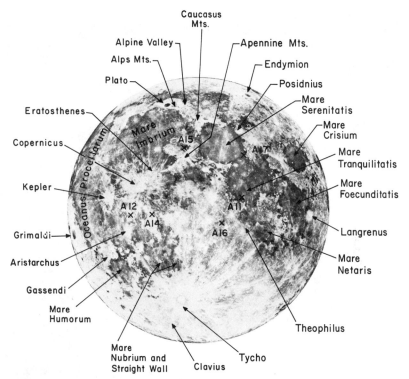

Fig. 84. Lunar features identified on reproduction of Fig. 83. The Apollo landing sites are indicated by the letter *A* and number. (By Joseph F. Singarella.)

tory to enjoy this privilege and to record his observations for posterity. Even with his small telescope he could detect the mountains, the craters, and the great dark areas that make up the features of the "man in the Moon." To him the dark areas looked like great seas of water, hence he called them *maria,* the Latin term for seas (singular *mare,* with the accent on the first syllable).

In Fig. 83, where the Moon is full, the maria can be seen to good advantage. A few of them and other conspicuous lunar features are identified in Fig. 84. The somewhat whimsical Latin names chosen for the maria cannot be explained on any rational basis, although Tranquilitatis, Serenitatis, and Frigoris seem appropriate enough. These maria are, of course, not seas but great plains, fairly flat except for the curvature of the surface, and devoid of both air and moisture.

Fig. 85. Mare Imbrium; Moon at third quarter. (Photograph by the Mount Wilson and Palomar Observatories.)

Mare Imbrium (Sea of Showers) and Mare Serenitatis, in the upper center of Fig. 83 are very large and nearly circular in shape, in so far as they are clearly outlined. The greatest diameter of Imbrium is over 1100 kilometers, and of Serenitatis 700 kilometers. A close-up of a part of Mare Imbrium is shown in Fig. 85. The magnificent mountain range outlining the lower right portion of the mare is known as the Apennines. These mountains rise some 6 kilometers above the level of the plain, a height that fully justifies the plagiarism in their name. The perspective given by the shadows in Fig. 85 reveals that the Apennines are great peaks rising sharply from the floor of the mare but sloping away gradually toward the outside. Deep valleys and cuts are numerous. The mountain range presents the appearance of models of terrestrial ranges, where valleys have been worn by the erosion of water—but no water exists on the Moon and never has, as the Apollo Moon rocks show.

Paralleling the inner edge of the Apennines can be seen a long somewhat crooked furrow, or *rill*. Several hundred rills have been found on the Moon. They are ditchlike depressions, hundreds of meters deep and extending for tens of kilometers along the Moon's surface. Since the rills usually do not show tributaries, as they should if they were formed by erosion, and since their walls are not raised above the surrounding "moonscape," they are most rationally explained as cracks. During cooling it is likely that the Moon's surface split open in places. Long or deep cracks may also have been filled in or overflowed by subsurface molten material, and show as a different type of marking. The long low ridge in the lower center of Fig. 85 appears to be an extension of the rill system already noted. A great crack that once opened at the base of the Apennines may now show as a rill along a part of its length and as a ridge in another part, while in between it may have been entirely covered by the flow of molten material.

The more scattered mountains in the upper right of Fig. 85 are the Alps. Their most conspicuous feature is the Alpine Valley, a giant cut through the center of the chain. The valley is 10 kilometers wide at its broadest portion and 120 kilometers long, with a level floor. The magnificent photograph of the Alpine Valley region by the U.S. lunar program (Fig. 86) shows a cen-

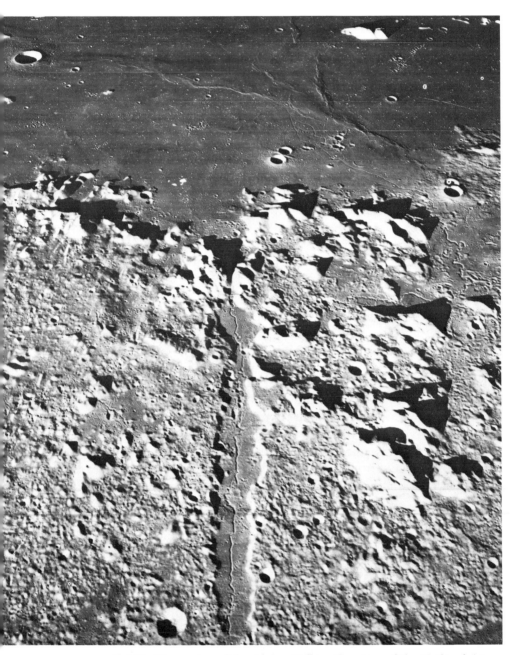

Fig. 86. The region of the Alpine Valley. (Courtesy of the National Aeronautics and Space Administration.)

tral rill in the valley, many other rills, and the irregular nature of the mountain range.

The rugged appearance of the Mare Imbrium circumference in Fig. 85 as compared to the general dullness of Fig. 83 is not produced by increased photographic contrast in the second reproduction, nor by an effect of enlargement. In the first view the Moon is full, and the Sun is shining directly down on it; hence the shadows are eliminated. In the second view the Sun is shining from the left, casting long shadows across our line of sight. Because of the curvature of the Moon, the extreme right edge of Fig. 85 is in darkness, except for the high mountain peaks. Along the *terminator,* between darkness and light, the rugged features of the Moon show to the best advantage. The great shadows betray the irregularities that may be invisible when the Sun shines overhead. Because of this effect, the Moon can be best observed when near the first or third quarter. The Sun's rays along the terminator are nearly perpendicular to our line of sight. At full moon, we can distinguish only the light and dark areas; the irregularities are lost.

The shadows serve a very useful purpose in providing an accurate measure of the heights of the lunar features. In Fig. 87a the isolated mountain peak Piton in the upper right area of

Fig. 87a. Piton, an isolated lunar peak in Mare Imbrium; Moon at third quarter. (Section of a photograph by the Lick Observatory.)

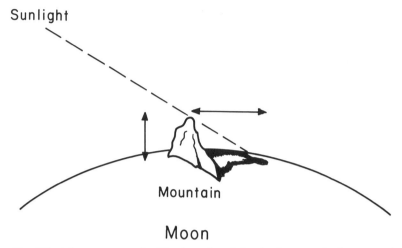

Fig. 87*b*. A lunar mountain. The lengths of the shadows cast by lunar markings enable astronomers to measure the heights. Compare with Fig. 87*a*.

Mare Imbrium is even further enlarged to accentuate the
shadow. The length of such a shadow can be measured, and the
angle of the Sun's rays can be calculated from the phase of the
Moon and from the known position of the mountain. Figure
87b illustrates the geometry of solving for the height. The calculations are straightforward, but somewhat involved because
of the several angles that must be considered. Although radar
elevation measures are important, this shadow method still
serves well in the space age.

Craters abound almost everywhere on the Moon; they become conspicuous when seen near the terminator. The varied
character of the craters is apparent in the region of Mare Imbrium, where they stand alone in the open plain. Some seem
smooth and flat within, while others show one or more central
peaks, often perforated by smaller craters. Crater walls barely
rise above the plain here and there, some of them only partially
complete. Close examination reveals that the Moon's surface is
covered with an almost unlimited number of small craters (cra-
terlets and crater pits), as though the Moon had been peppered
with buckshot. The smallest craters identifiable in the Earth-
based photographs are usually a kilometer or two in diameter.

The craters can be classified into several types according to
their forms. Since fine distinctions can be extended intermina-

Fig. 88. Lunar bad lands. The southern area of the Moon at third quarter.
(Photograph by the Mount Wilson and Palomar Observatories.)

bly and since the definitions are not always concise or uniform, it is perhaps better not to stress the type names. Many craters possess interior plains as flat as the maria and mountain walls that rise abruptly to define the edges. These are known as bulwark plains, bulwarked plains, or walled plains. The level of the plain may lie above or below the general level outside. The largest clearcut crater on our side of the Moon, Clavius, is a bulwarked plain, with a maximum diameter of 235 kilometers from opposing mountain summits. Clavius can be seen near the bottom center of Fig. 88. The curvature of the Moon's surface is sufficient to hide the 6.6 kilometer mountain walls from an observer standing in the center of the plain.

The crater Tycho, seen slightly below the center of Fig. 88, represents a somewhat different type of crater formation, often called ring mountains. Only a small fraction of the basin is flat, the crater being more nearly saucer-shaped. The inner slope of the mountain rim is itself ringed, somewhat as though laminated or terraced. These ring-mountain craters are also fairly perfect in form, almost circular, and are rarely encroached upon by lesser craters or other deformations. The ring mountains thus show evidence of having been formed later in the Moon's history than the bulwarked plains, which bear the scars of subsequent tribulations. Other fine examples of the ring mountains are Eratosthenes and Copernicus (Fig. 85, upper right, and Figs. 89 and 90). These are all huge impact craters.

The rougher areas of the Moon (Fig. 88) are completely covered with a wild hodgepodge of craters within craters and craters upon craters. They all appear to have been formed in an entirely hit-or-miss fashion, the newer ones evolving with a complete disregard for all that were there before. Sections of a wall may stand after an old crater has been partly demolished by a new one, and this, in turn, may be pock-marked by smaller craters still more recent.

Across these rugged areas of the Moon and across the extensive plains run great systems of *rays,* the light-colored streaks that show so conspicuously when the Moon is full (Fig. 83) but almost disappear at the partial phases (Fig. 82). A most notable system centers on the crater Tycho (diameter 87 kilometers), from which the rays can be traced almost around the Moon. In Fig. 88 the rays are barely discernible, and Tycho has become

Fig. 89. Copernicus. Moon past third quarter. Compare with Figs. 83, 90, and 91. (Photograph by the 100-inch reflector of the Mount Wilson and Palomar Observatories.)

just one of many craters, not the most conspicuous of all. The rays cast no shadows and can be detected only by their lighter coloring. They are broken neither by mountains nor by any other features of the lunar topography. They are certainly splash marks from large, relatively new craters.

Fig. 90. Part of the Copernicus area, photographed by U.S. Lunar Orbiter 4. (Courtesy of the National Aeronautics and Space Administration.)

Note the complex structures about the craters Copernicus and Kepler in Fig. 83. The light color of the rays is shared by the craters with which they are associated, and also distinguishes the rims of a large number of craters. At first glance the region about Copernicus in Fig. 83, when the Moon is full, can hardly be identified with the area about Copernicus in Fig. 89, at a later phase. The bright crater rims in the latter figure, however, can

Fig. 91. Copernicus crater. Diameter 93 kilometers. (Courtesy of the National Aeronautics and Space Administration.)

soon be detected in the former, while some of the larger craters with dull edges have faded to invisibility. A careful comparison of the two photographs is most instructive. The second largest crater in Fig. 89, Eratosthenes, can hardly be found in Fig. 83, although it is very similar to Copernicus in type. Copernicus, Eratosthenes, and Kepler are 93, 58, and 32 kilometers in diameter, respectively.

To the right of Copernicus in Figs. 89 and 90 are several crater chains, forming a pattern. Such crater chains are not uncommon on the Moon's surface. They are especially noticeable when viewed with a high magnification under good seeing conditions. The highly magnified section of regions near Copernicus in Fig. 90 shows formations that appear to be *volcanic domes* and *volcanic sinks*. Some of the domes have central craters and resemble terrestrial volcanoes of the Vesuvius type.

Long serpentine ridges can be seen in the flat plains of Mare Imbrium (Fig. 85). Some must cover older markings and many are the edges of lava flows.

Another interesting lunar formation is shown in Fig. 92, the Straight Wall (or the Railway) some 113 kilometers in length, which appears also in Fig. 88. In this picture, the sun is shining from the left and the Straight Wall shows as a *white* line (to its left, nearly parallel to it is a shorter curved rill). The photograph of Fig. 92 (*left*) was made just after first quarter so that the Sun shines from the right. The Straight Wall now shows as a *dark* line, which proves that the marking is a long straight cliff or wall facing toward the left. Measurement shows that much of it is elevated some 660 meters above the plain. It is not as steep as we might suspect, however. J. Ashbrook finds that its slope is not more than 40°; it is clearly a rock *fault,* where one edge has risen above the other. Violent moonquakes, in the dim and distant past, were probably associated with this and other walls visible on the Moon. Similar fault markings, frequent but smaller on the Earth, are the foci of earthquakes. On the Moon, large systems of parallel cracks and faults can be traced, indicating that the surface has been strained and distorted by internal forces. Figure 92 (*right*) presents a lunar *graben,* where the land has been stretched and the crack filled.

The other side of the Moon no longer remains a mystery. On October 4, 1959, the U.S.S.R. first obtained images of the Moon

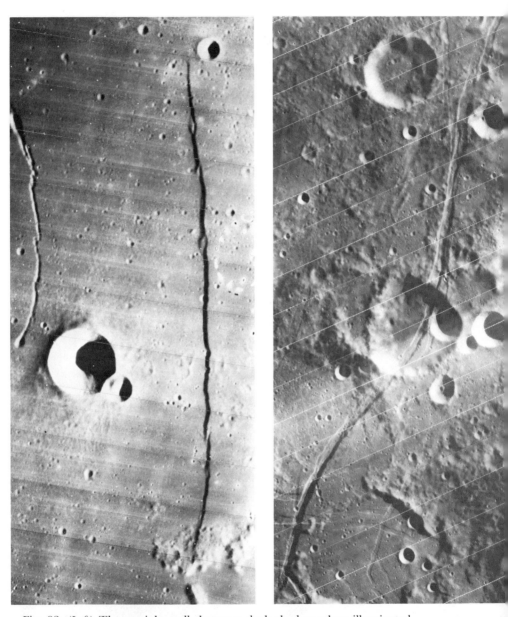

Fig. 92. (*Left*) The straight wall throws a dark shadow when illuminated from the east. Compare Fig. 88. Width of view: 62 kilometers. (*Right*) This portion of Rima Sirsalis in Oceanus Procellarium near the crater Grimaldi is a lunar graben. Width of view: 90 kilometers. (Courtesy National Aeronautics and Space Administration.)

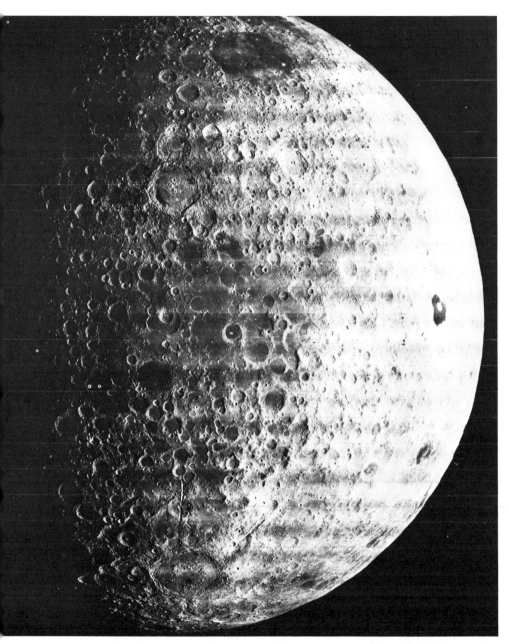

Fig. 93. Part of the far side of the Moon, photographed by U.S. Lunar Orbiter 4, centered near the eastern limb as seen from the Earth. (Courtesy of
the National Aeronautics and Space Administration.)

Fig. 94. Mare Moscoviense is the largest mare on the Moon's far side, about 350 kilometers across. By Lunar Orbiter (Courtesy of the National Aeronautics and Space Administration.)

from a "cosmic rocket" passing beyond the Moon. Since then the U.S. Orbiter spacecraft and Apollo missions by the National Aeronautics and Space Administration have surveyed almost the entire lunar surface with resulting photographs of extraordinary quality and detail. Figures 93 and 94 show the far side of the Moon and a view of its largest mare. Note the 240 kilometer north–south trough in the lower center of Fig. 93, much like the Alpine Valley but longer.

The major scientific surprise given us by the far side of the Moon is, indeed, the lack of maria. Whereas the near side is about half covered with maria, the far side shows only one relatively small one! The speculations aroused by this observation are still not satisfied by any completely acceptable scientific explanation. Many astronomers favor John A. Wood's theory that the Earth's gravity caused the near side of the Moon to be more thoroughly peppered with large asteroidal type bodies than the far side.

There are so many interesting formations on the Moon's surface—individual craters with unusual structures, peculiar rills, rays, maria, mountains, and cliffs—that descriptions could be continued indefinitely. The reader, however, may wish to do some exploring himself, by means of these photographs or with almost any telescope that can be firmly supported. The huge collections of photographs obtained by the National Aeronautics and Space Administration's Ranger, Surveyor, Orbiter, and Apollo spacecraft offer a lifetime of detailed study of the Moon. Larger-scale charts can be obtained easily by anyone who wishes to learn the proper names of individual lunar formations. Some 5000 markings are officially designated by the International Astronomical Union, and many thousands more have been plotted by assiduous selenographers. The close-up spacecraft photographs are now taxing the official map makers with a job comparable to that of mapping the Earth itself.

The more serious observer can join the ranks of those dedicated amateurs who have contributed so much to solar-system astronomy. In most countries he can find groups who are engaged in patrolling the lunar and planetary surfaces for changes that the professional astronomer will probably miss. Even with the modern acceleration of observational astronomy, there are far too few professionals for the task at the telescope.

9

The Nature of the Moon

Mankind's greatest triumph of exploration is certainly the landing of men on the Moon and their safe return to Earth. The Apollo landings were the culmination of a program of "hard" and "soft" landings beginning with the U.S.S.R. hard landing in 1959. The wealth of photographs of the Moon, the close-up studies, the direct exploration, the subtle analysis of lunar material returned by Apollo astronauts and by the U.S.S.R. unmanned vehicles, the seismographs radioing back records of moonquakes, the pitting of surfaces on Surveyor I after three years on the lunar surface, the laser echoes from the corner reflector left at Apollo landing sites, and the high-energy atoms, ions, and electrons in space around the Moon, coupled with radio, radar, infrared, and ultraviolet measures from Earth and space, combine to provide a breathtaking store of information about the Moon. We now know more about the Moon, and perhaps Mars, internally and externally, than we knew about the Earth at the turn of this century. Many old questions are clearly

answered while the new questions are far more subtle and penetrating than the old. This and the following chapters will be
drawn from this storehouse of knowledge with only partial reference to the detailed sources, in an attempt to describe these
celestial objects as we know them today and to discuss how they
may have come into existence.

That the Moon's surface is a sublime desolation is fully confirmed by the Apollo astronauts. The lunar plains are more barren than rocky deserts. The lunar mountains are more austere
than terrestrial peaks above the timber line. Lava beds of extinct
volcanoes on Earth are more inviting than the lunar craters
(Fig. 95). There is no weather on the Moon. Where there is no
air there can be no clouds, no rain, no sound. Within a dark
lunar cave there would be eternal silence and inaction excepting
weak moonquakes. A spider web across a dim recess in such a
cave would remain perfect and unchanged for a million years.

There are no colors in the Moon's sky, only blackness and
stars during the bitter night, 2 weeks in length, and then the
glaring Sun during the equally long day. But there is danger to
man on the Moon's surface. Meteoritic dust may puncture his
space suit or his pressurized dwelling and may splash lunar material at rifle speeds to accomplish the same end. Fortunately
our Apollo astronauts escaped these hazards. But cosmic rays
and high-energy particles from the Sun struck them directly because there is no buffering atmosphere.

If any permanent changes have occurred in the lunar land-
scape during the centuries of telescopic observation, the
changes are too small or uncertain for the observers to be able
to agree upon their reality. None of the lunar landing areas are
detectable from Earth except for laser reflections from the corner reflectors and radio transmission from the lunar seismographs.

Lunar exploration places the lunar atmosphere at practically
a perfect vacuum by laboratory standards. The number of
atoms or molecules per unit volume is a fraction of a million
millionth of our sea-level atmosphere and are mostly contributed by the solar wind.

We should expect the Moon to be devoid of an appreciable
atmosphere because of the smallness of its mass. Its surface
gravity is insufficient to prevent the molecules of an atmosphere

Fig. 95. Crater Alphonsus, photographed by U.S. Ranger 9. See Fig. 88, second crater from top, right center. (Courtesy of the National Aeronautics and Space Administration.)

from being hurled into empty space. Any body, large or small, moving away from the Moon's surface with a speed in excess of 2.4 kilometers per second will continue to recede indefinitely, completely out of the gravitational control of the Moon. This critical velocity of escape is only slightly greater than the *average* speed of a hydrogen molecule in a gas at ordinary temperatures. Since some of the molecules must always move faster than the average, a hydrogen atmosphere would dissipate from the Moon almost instantly. The dissipation of oxygen or nitrogen would be very much slower because the molecules are heavier than those of hydrogen. In a short time astronomically, however, the Moon will lose any atmosphere it might once acquire. The Moon now gains some atmosphere from interplanetary space, mostly fom the gas clouds shot out by the Sun and the gas lost from the solar corona. This extraordinarily tenuous atmosphere is continuously lost, while the high energy of the incoming atoms knocks away any heavier atoms that may be oozing out of the Moon's interior.

Among the active forces that are probably most important today in altering the Moon's surface are human beings, then meteoritic impacts and the high-energy particles, particularly x-rays and cosmic rays, that penetrate the top surface layers. None of these natural forces may produce much change during a human lifetime, but over hundreds of millions of years they have "gardened" the Moon's top surface layer, or *regolith.* We will probably make striking alterations in the Moon's appearance, but that is for the future. In February 1967 the U.S. Lunar Orbiter 3 photographed the Surveyor 1 spacecraft that soft-landed, June 2, 1966, on a typical "smooth" mare area in Oceanus Procellarum. The picture shows the white spot of the spacecraft and its shadow some 10 meters in length. The crash landing of the U.S. Ranger 8 spacecraft produced a crater some 13 meters in diameter, as identified in an Orbiter photograph. Several other crash landings on the Moon undoubtedly produced similar craters.

Each meteoroid striking the Moon produces a miniature explosion, throwing rocky and meteoritic materials out in all directions. At a speed of 16 kilometers per second the particle can impart a velocity greater than 2.4 kilometers per second, the velocity of escape from the Moon, to a fraction of its own mass of

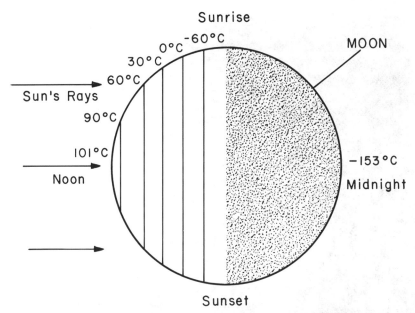

Fig. 96. Temperatures on the Moon. (From measures by E. Pettit and S. B. Nicholson.)

material. Since there is no atmosphere to resist the loss, such material leaves the Moon's gravitational field, although much may be recaptured from Earth orbits.

The alternations in temperature from the lunar noon to midnight are extreme, more than 250°C (Fig. 96), but they take place slowly. The *exfoliation* or flaking away of the surfaces of rocks on the Earth is due chiefly to the expansion of absorbed moisture on freezing. Because of the absence of water on the Moon, exfoliation could result only from thermal expansion and contraction. However, the very gradual changes in temperature on the Moon allow the rocks sufficient time to adjust their internal temperatures so that exfoliation by pure expansion and contraction is very slow.

Although the enormous temperature changes from day to night occur very slowly, large changes are observed during an eclipse of the Moon. E. Pettit and S. B. Nicholson of the Mount Wilson Observatory first measured the Moon's temperature by its infrared (or heat) radiation through the course of a lunar eclipse. The temperature fell from 71°C to −79°C in about an

Fig. 97. First radar contact with the Moon. At a wavelength of about 1.5 meters in January 1946 this antenna at the U.S. Signal Corps Engineering Laboratory, Fort Monmouth, N.J., first bounced radar pulses off the Moon. (Official U.S. Army photograph.)

hour. Such a quick alternation in temperature may be more active in exfoliating newly exposed lunar rocks, but only on the near side of the Moon.

Had the Earth been unable to retain an atmosphere, its surface would probably be similar to that of the Moon, barren and probably rougher. Air and water, however, have made possible the surface that we know and have led to the evolution of living forms. No life could develop on the surface of the Moon, although conceivably there may have been a time during its formation when there was appreciable atmosphere. The lunar rocks returned by the U.S. Apollo astronauts and the U.S.S.R. Luna landers, as mentioned earlier, show no evidence of *ever* having been exposed to water.

Radio telescopes have added enormously to our knowledge of the Moon's distance (Fig. 97) and temperature. Active radar pulses can be timed in transit to give the distance and also the roughness of the lunar surface. Passive radio receivers, on the other hand, can measure the radiation emitted by a surface and thus determine its temperature. All ordinary materials radiate more energy at all frequencies when they are heated. Radio waves may pass through the porous upper layers to give an indication of the temperature beneath. At wavelengths of 30 centimeters or longer there is practically no temperature change observed during the month or even during lunar eclipses, while at successively shorter wavelengths, nearer to the infrared, temperature changes are increasingly greater. A meter or so below the Moon's surface on the equator the temperature appears to remain constant at $-50°C$ or perhaps somewhat colder.

These combined results show that the Moon's top surface layers are made of very porous material, an excellent insulator. This insulating layer holds little heat and thus can respond quickly to changes in solar radiation, at sunset, sunrise, or eclipses of the Moon. Infrared measures over the near surface of the Moon show that many areas remain warmer during eclipses and in the lunar "evening" than does the average lunar surface. Figure 98 shows a map of these regions, which comprise almost all the new (white ray) craters such as Tycho and some of the maria. The lunar highlands are relatively free of such "hot" spots except for new craters.

We can only conclude that more recently disturbed areas of the Moon have not had time to develop so thick an insulating surface as the older areas. Thus the heat accumulates during the lunar day in rocks or good conducting materials, to be radiated away during the night. Recall the warmth from a wall in the early evening after it has been exposed to the Sun during

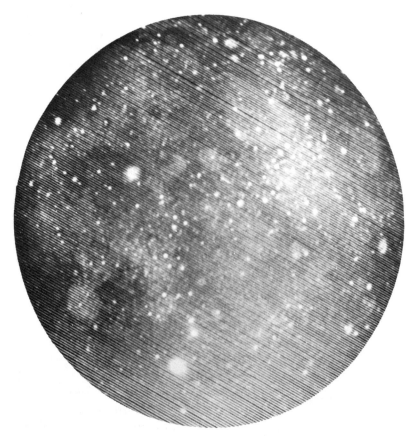

Fig. 98. Infrared picture of Moon during eclipse shows white spots that cool more slowly than remainder of surface. Compare Figs. 83 and 84 to identify major ray craters and certain maria. (Courtesy of R. W. Shorthill and J. M. Saari, Boeing Aircraft Co.)

the afternoon. Such a material has technically a high *heat inertia* as compared, for example, to cork. The observations do not indicate that any areas on the Moon actually radiate heat from an internal source, say of a volcanic nature.

Since dust in a vacuum is an excellent insulator to heat, the rapid temperature changes during eclipses have long suggested that the upper layer of the Moon's surface should be dust. The great increase of the Moon's brightness near the full phase indicates that the surface is extraordinarily rough at dimensions of less than a millimeter, a fact which has been confirmed by the astronauts. The total light of the Moon doubles in the 2 days before full Moon and the surface appears nearly uniformly bright to the edge at full. But long-wavelength radar observations of the Moon show that its surface is fairly smooth, with about half of its area tilted at more than 8° from the spherical. From limited data the maria appear smoother than the craters and mountains. At a wavelength of 8 millimeters, radar measurements made at the Lincoln Laboratory of the Massachusetts Institute of Technology show the Moon's surface to be much rougher than when measured at longer wavelengths. The central area of reflection is much larger. In visual and infrared light, however, the full Moon appears almost uniformly illuminated.

The U.S.S.R. Luna 9 landed on January 31, 1966, to give us our first close-up view of the Moon (Fig. 99). Since then the NASA Surveyors and the Apollo manned landings with sample returns provide complete information about a number of areas on the Moon. The top material is mostly very fine grained, the

Fig. 99. The first close-up of the Moon. From the U.S.S.R. Luna 9. The largest nearby rock is about 30 centimeters in diameter.

Fig. 100. Footprint of Neil A. Armstrong or Edwin E. Aldrin, Jr., Apollo 11, July 24, 1969. They made the first human footprints on the Moon. (Courtesy of the National Aeronautics and Space Administration.)

grains being on the order of a few thousandths of a centimeter in dimension, interspersed with occasional rocks from pebble-size upwards. The bearing strength is weak, so that astronauts walking on the surface leave clear-cut footprints a centimeter or so deep (Fig. 100). If not disturbed by man they will remain visible for a million years.

Cuts in the surface leave rather vertical walls, indicating that the material has internal strength, but is weak, perhaps stronger than freshly plowed earth. When gouged, the surface tends to crack slowly in small blocks, confirming that it is a somewhat crunchy material. In other words it is not very compressible under moderate pressures, suggesting an internal structure of some coherence. The surface density is a bit more than that of water. It is porous and becomes denser and stronger at depths over a few centimeters.

Occasional rocks protrude or lie on the surface. These appear to have been thrown from crater explosions. Note the boulder tracks in Fig. 101. Large-scale steep slopes on the Moon are rare except in craters. We have noted that the Straight Wall has a typical slope of only 40°; the average slopes of the inner walls of the crater Tycho are only about 17°. The NASA Ranger spacecraft pictures of the Moon show that only one percent of the

Fig. 101. Hundreds of boulder tracks like this one have been studied. These are on the inner wall of a 45-kilometer crater. The larger track is about 25 meters wide. (Courtesy of the National Aeronautics and Space Administration.)

surface is inclined more than 13°. The great shadows seen near the terminator deceive us as to the ruggedness of lunar features. Mount Piton, for example (Fig. 87a), rising more than 2 kilometers above the mare floor, stretches out more than 20 kilometers at its base and has a level top.

One major puzzle of the lunar surface is finally clarified. Why is the surface so very dark? Fine dust from broken rocks is a much better light reflector than the Moon's surface. The trench made by the shovel of Surveyor 3 shows that the very top layer (less than a millimeter thick) is slightly lighter in color than the deeper material. But why is the deeper material so very dark and why are some of the rocks so much whiter? For a while it was thought that the high-energy ions of the solar wind would slowly darken the surface.

Extensive laboratory studies of lunar samples show that the grains are covered with thin coatings that absorb light to make them dark. Two processes working together in the near vacuum of the Moon's surface are responsible: solar wind particles do indeed "sputter" off atoms, while micro-meteor impacts add vapor; a part of both strike the surfaces of the grains. Heavier atoms such as iron and titanium tend to stick better than lighter atoms such as oxygen or silicon. Thus the lunar grains slowly develop extremely thin dark coats. The "gardening" effect of meteor impacts turns over the regolith currently at rates of the order of a centimeter in ten million years, and at much higher rates early in the Moon's history. Thus the darkening extends to grains that are well buried. As an example of the gross effects of "gardening," consider the measures from the Apollo 17 drill sample. The material is fine grained to a depth of 18 centimeters, where a coarse-grained layer appears. The solar flare implantations indicate a cratering event around 100 million years ago, which covered the region with coarse-grained rubble. Being on a slope, the coarse-grained material was subsequently covered over to a depth of perhaps 25 centimeters by material knocked down the hillside. The Apollo core was made in a shallow depression, which indicates that a very small crater formed on top much more recently and has been partially filled.

Some small craters show an inner terrace or even two, suggesting that a harder stratum or strata lie a few meters to tens of meters below the surface. Such layering of the lava flows is evident in several of the maria regions that are well studied. In addition, in these regions of the maria the craters become *saturated* below diameters of perhaps fifty to a few hundred feet. That is, new craters have overlapped and covered up old craters to produce a statistically stable distribution of craters. The addition of new craters would not change the general appearance of the surface.

The close-up lunar pictures all show that some of the sharply defined "new" craters have white rims and some have rims of

Fig. 102. A 60-meter-diameter crater strewn with meter-sized boulders in Oceanus Procellarum. (Courtesy of the National Aeronautics and Space Administration.)

the same shade as their surrounding. Hyperballistic experts suspect that this difference arises from the velocity of meteoritic impact. At low velocities the meteorite causes a subsonic explosion and the broken material is merely pushed out. But at velocities of 10 kilometers per second or higher, typical of meteorites from space, much of the material is pulverized into very fine dust and thus gives a white rim. Figure 102 shows a very rocky crater wall and surroundings, probably caused by a low-velocity impact.

The huge rays from the great new craters such as Tycho cannot, however, be explained by white dust alone. The U.S. Ranger 7 pictures have confirmed Kuiper's telescopic observation that the rays are rough and rocky. White rocks, such as appear in the close-up pictures, could cover the surface of the rays

Fig. 103. The crater Wargentin. The two largest craters are Schickard (*lower*) and Phocylides (*upper*), with Wargentin, the filled crater, between them to the left. (Photograph by the 120-inch reflector of the Lick Observatory.)

sufficiently to keep them relatively white for long periods of time, until they were slowly covered by debris thrown from more distant parts of the Moon. Their increase in relative brightness at full Moon, however, requires further explanation. Possibly their very roughness produces the effect. The "shadow" of an aircraft at great altitudes becomes a bright spot on ordinary ground surfaces although it is a dark shadow on water. On foliage, for example, (see Fig. 181) light rays opposite the Sun can retrace their original path to the eye. At other angles the light is dimmed in finding a new path. Thus if the lunar bright rays are rougher on a large scale than the surface of the Moon on a microscopic scale, they then show their true higher albedo near the full phase.

For many years before the space age there was great speculation about the origin of the Moon's surface features. Has the Moon duplicated the volcanic or plutonic activity of the Earth with great volcanic craters and lava flows? Or was it always rather inactive, with meteoritic impacts being the great surface molding agent? Was the Moon ever molten or partially molten?

We look to the Moon's surface itself for visual evidence concerning melting and volcanic action. We note in many areas of the Moon, particularly on the far side, a chaotic maze of crater upon crater, constituting the rough "highlands" area of the Moon. In contrast, the surfaces of the maria appear relatively smooth and contain many fewer intermediate-sized craters per unit area. There, however, we find abundant evidence for volcanic or plutonic activity: the crater chains near the great crater Copernicus, mentioned earlier (Fig. 89), and in a region east of Copernicus where volcanic sinks and domes also appear (Fig. 90). The remarkable crater Wargentin (Fig. 103) shows clearly that lava was actually forced to the top of the crater walls, leaving it essentially full, or that regions nearby later subsided from a common level. Volcanos generally cause a mountain of material to rise about a central vent, as in Vesuvius, the volcanos of the Hawaiian Islands, and many others (Fig. 104). Lunar evidence for volcanos of these dimensions can be seen in Figs. 90 and 91, but all of these clear-cut cases of volcanos and volcanic domes are relatively small in size.

Color photographs of the Moon with extremely accentuated color contrast by E. Whitaker show patchy irregular patterns

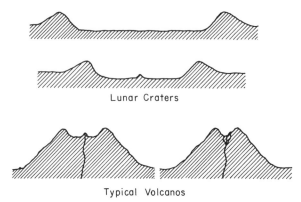

Lunar Craters

Typical Volcanos

Fig. 104. Craters and volcanoes. These schematic cross sections indicate the basic differences between lunar craters and volcanoes of the Vesuvius type.

over the maria basins. The edges of these patterns match the edges of very low-lying plateau areas and scarps, clearly the edges of lava flows. Lunar Orbiter pictures confirm these lava flows as in Fig. 105.

Cracking, melting, slumping, filling, ridging, and plutonic activity of many kinds are apparent almost universally not only on the maria but within the filled craters and in flat regions between lunar mountains. For example, the U.S. Ranger 9 views within the crater Alphonsus (Fig. 106) show that the rill system and most of the small craters must have arisen from a geological type of activity. Dark halo and slump craters are abundant. The floor of the crater Aristarchus (Fig. 107) shows a somewhat chaotic surface, which resembles the floors of some Hawaiian volcanos. Even more chaotic is the floor of the great 85-kilometer ray crater Tycho, in the lunar highlands (Fig. 108). Because the great impact created an original crater tens of kilometers deep, we are seeing here the rsults of fall-back, melting, and secondary igneous activity on a mad scale. The U.S. Orbiter 2 view of Oceanus Procellarum in the neighborhood of Crater Marius (Fig. 109) shows not only serpentine ridges that certainly were formed by lava or magma flows but also volcanic domes of diameters 3 to 15 kilometers and heights 300 to 450 meters, some with visible craters. The remarkable complexity of the maria surfaces seen close up, as compared to their drabness in Earth-

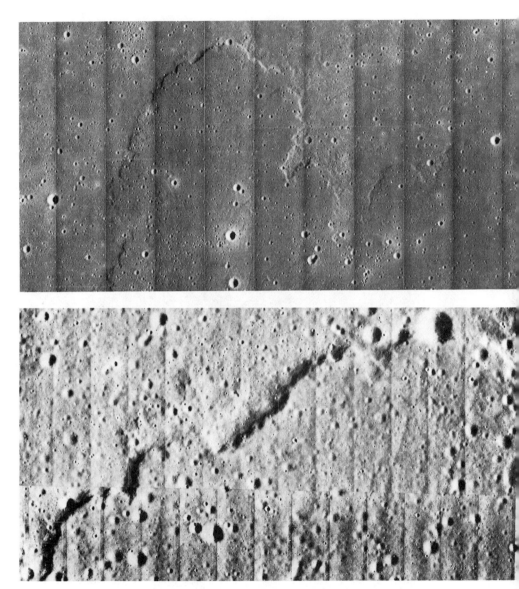

Fig. 105. Lava flow in Mare Imbrium. (*Top*) Picture is 27 kilometers wide, North to left. (*Bottom*) Close-up of left center. Picture is 4.5 kilometers wide. (Courtesy of the National Aeronautics and Space Administration.)

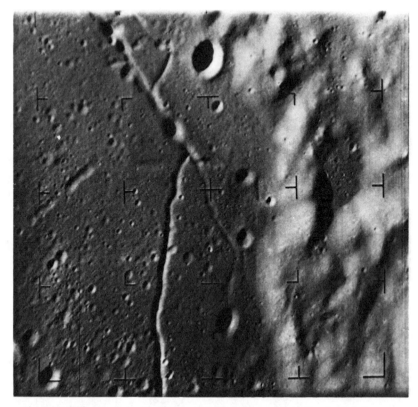

Fig. 106. Rill system inside the crater Alphonsus. Note the dark halo crater, bottom center, which may have been darkened by gas ejection. (Courtesy of the National Aeronautics and Space Administration.)

based photographs, attests to the plutonic activity of the Moon in her youth.

The larger craters on the Moon are all or mostly meteoritic. Ralph B. Baldwin in 1949 amassed such an array of evidence for the meteoritic theory that very few proponents of the volcanic theory persisted to the space age. The number of new craters on the maria fits well with the expected influx of large meteorites over 2 to 4×10^9 years. There was the old counter argument that, if such large meteoritic bodies made the great craters on the Moon, the Earth should carry even greater scars because of its increased gravity and the consequent greater velocity of fall of similar bodies to the Earth. The Earth *has* many scars of this type, in the form of crypto-volcanos, not to men-

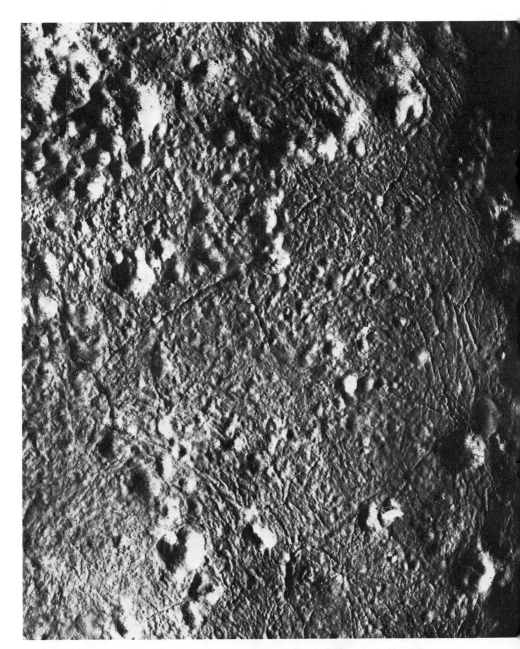

Fig. 107. Section of the floor of the Crater Aristarchus. Width of picture 5.7 kilometers. (Courtesy of the National Aeronautics and Space Administration.)

Fig. 108. A 5-kilometer-wide section showing the rough interior of the ray Crater Tycho. (Courtesy of the National Aeronautics and Space Administration.)

Fig. 109. A view of a lunar mare in Oceanus Procelarum near the crater Marius, upper right, which is 40 kilometers in diameter and 1.6 kilometers deep. (Courtesy of the National Aeronautics and Space Administration.)

tion the recognizable recent meteor craters such as the Barringer Crater in Arizona and larger ones well demonstrated by C. S. Beals in Canada (Fig. 53). Geological forces have, of course, filled large craters of the past, tilted them, eroded them away, and in large measure destroyed their record. A striking example is the Vredefort Dom in South Africa, which was originally some 50 kilometers in diameter. A discussion of these fossil meteoritic craters on the Earth is found in *Between the Planets* by F. G. Watson.

We know from our experience with nuclear weapons that large craters can be formed by explosions. High-velocity mete-

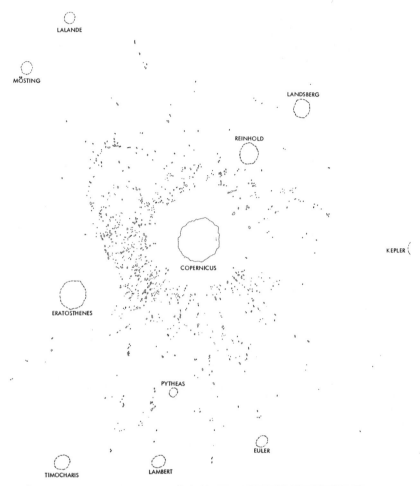

Fig. 110. Secondary impact craters in the area of Copernicus; compare with Fig. 89 inverted. (After E. M. Shoemaker.)

orites can produce similar effects on *any* scale. E. M. Shoemaker's diagram in Fig. 110 shows the impact craters in the area around Copernicus (compare with Figs. 89 and 90). These tiny craters, still of the order of a kilometer in diameter, have been produced by the material thrown from this explosion. No volcanic activity that we know of could produce such large explosive action as to throw out masses of millions of tons of material to distances of 200 kilometers and much farther. On the other hand, craters of volcanic type do appear in parts of this same area.

Lunar craters above 10–30 kilometers in diameter usually show central peaks and wall terraces. Physical theories and computer modeling of huge impacts explain these features. The impacting body at a speed much above the velocity of sound in rocks buries itself several diameters, creating a shock wave which compresses and heats the target rocks. They in turn flow like mud, splashing out to form a deep temporary crater. The slushy rims and walls slump to form terraces and to refill much of the crater. The pressure rebound from the bottom combines with this infall to produce central uplifts and small mountains. Molten rock left inside seeps to the center, the source of mild volcanic activity. The remarkable U.S. Orbiter 3 view of the Copernicus crater basin (Fig. 111) and north walls shows the complexity of plutonic processes after the meteoritic explosion.

Let us now face the question whether the maria were formed primarily by volcanic action or were first triggered by the impacts of extraordinarily large meteorites. The latter solution was first suggested and defended late in the last century by the great geologist G. K. Gilbert (1843–1918), supported by Baldwin in 1949, and more recently by the space program.

An intense study of the lunar surface shows that at least one-half of the visible surface of the Moon exhibits markings that are linked to Mare Imbrium. Thousands of cracks and filled valleys point radially to a point near its center, and also systems of cracks, mountain chains, and other formations occur almost circularly symmetric about this point. Figure 112, due to Kuiper, shows this system on a projection normal to the Moon's surface so that the foreshortening of Mare Imbrium, as seen from the Earth, has been eliminated. He identified the remarkable inner basin surrounded by serpentine ridges marked in dark. This is a nearly square area symmetric about the presumed impact point. The Alpine Valley points directly toward this central point and the great mountain ranges around are quite symmetric with respect to it. Within these mountain ranges lies the outer basin in which there appear to be great blocks that represent either original lunar material not destroyed by the formation of the mare or material forced out into that region in the formation of the mare.

Most everyone now agrees that an enormous meteoritic body landed near this impact point perhaps with a large component

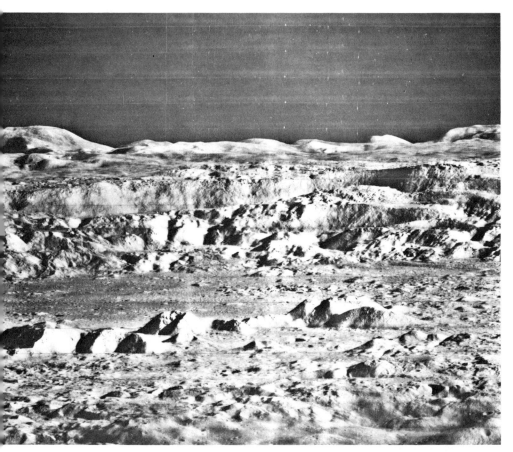

Fig. 111. View by U.S. Lunar Orbiter 3 of the Copernicus crater. The mountains just beyond the foreground can be seen as the central peaks in Figs. 89 and 90, looking from the south, that is, from the bottom of the figures. (Courtesy of the National Aeronautics and Space Administration.)

of its velocity toward the south (up in the photograph). Within the inner basin a huge crater was formed and materials were thrown out to extreme distances, far beyond the apparent center of the Moon as seen from the Earth. The great impact occurred during the interval when the Moon's outer layers were partially molten, so that the huge mass ejected from the central crater presented a tremendous load for the lunar crustal rocks to support. The present region between the outer basin and the inner edges of the mountain ranges failed to support the load and subsided. The inner edges of the Alpine Mountains, the

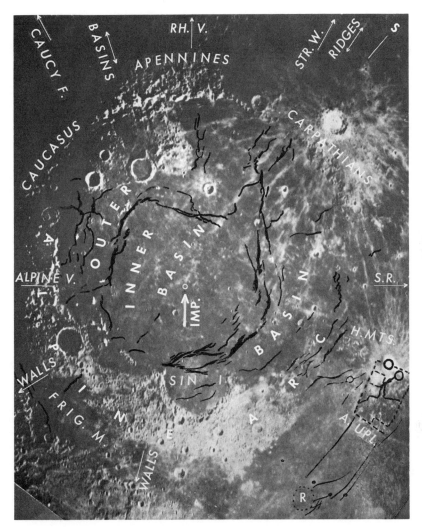

Fig. 112. Structures in Mare Imbrium as delineated by Gerard P. Kuiper. (South is up.)

Caucasus Mountains, the Apennines, and the Carpathians are thus slip faults formed as the inner and initially higher regions sank. Even though some of these mountains are more than 8 kilometers high, we note that lunar gravity is only one-sixth that of the Earth. Thus the highest mountains on the Moon require a supporting force equal only to mountains just over a kilometer high on the Earth. Hence, if only a small fraction of the upper layers of the Moon were molten, it seems reasonable that enough support could be found from underlying and undisturbed lighter rocks for such a broken and incomplete ring of mountains as now remains.

The region on the Moon near its apparent center, extending to the Apennine Mountains and westward toward the Haemus Mountains near Mare Tranquilitatis, appears muddy and poorly resolved even in the best observing conditions. It is difficult to explain this unusual appearance except by assuming that the area had been covered to a considerable depth by partially melted material thrown radially from Mare Imbrium. This is the basis of Kuiper's argument that the incoming meteoritic body was moving in a southerly direction.

The space studies of the Moon reveal an even more spectacular basin than Mare Imbrium. This feature, Mare Orientale (Fig. 113), has long been partially visible at the Moon's limb, but its true appearance could only be perceived from space. Its outer scarp, the Cordillera Mountains, is almost perfectly circular and about 1000 kilometers in diameter, rising some 6 kilometers above the plain, quite comparable to Mare Imbrium in diameter. The inner ring of Rook Mountains is over 600 kilometers in diameter. Material has been thrown beyond the Cordillera scarp for another 1000 kilometers, constituting a unique blanket of complicated flow structures largely radiating outward from the center, obliterating most of the earlier surface features.

For both Mare Imbrium and the Orientale basin, gigantic impacting meteorites or comet nuclei, some tens of kilometers in diameter, have exploded craters also tens of kilometers deep and with diameters of hundreds of kilometers. The subsequent healing process of magmatic readjustment proceeded more completely for Mare Imbrium. Mare Orientale appears to be more recent. The concentric ring structures of mountain scarps and collapsed surfaces have aroused a number of theoretical speculations. Do they result from some supersonic shock wave phenomenon of hyperballistic impact, or do they literally reflect stratification layers at various depths beneath the lunar surface? The former explanation is generally accepted today.

Final evidence that the great maria contain dense material, therefore frozen dense lava, comes from the orbiting spacecraft

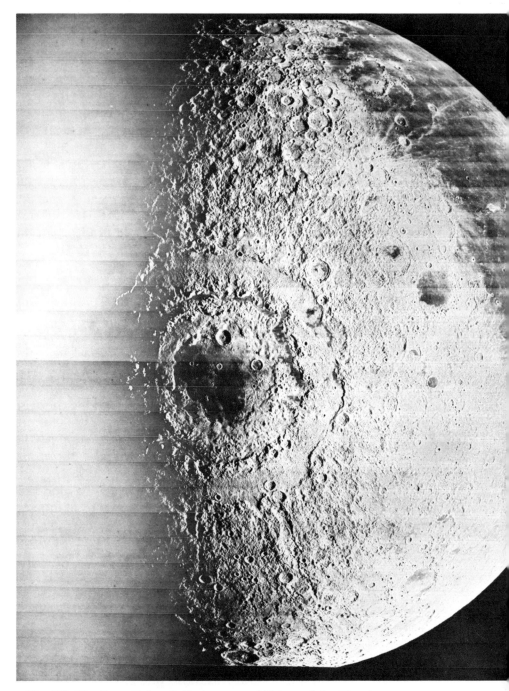

Fig. 113. The Great Orientale Basin centered 9° beyond the Moon's eastern limb 15° south of the equator, as viewed by U.S. Lunar Orbiter 4. Oceanus Procellarum appears in upper right. (Courtesy of the National Aeronautics and Space Administration.)

around the Moon. Their motions show that the gravity field above Mare Imbrium, for example, exceeds that above the highlands. The Apollo 17 laser experiment, on the other hand, shows that the surfaces of the maria and most of the near side of the Moon are nearer the center of the Moon than is the far side. In other words, the Moon is lopsided, the far side extending irregularly out from the center of mass on the average of about a kilometer while the near side is depressed by perhaps 2 kilometers. The maria themselves outline a nearly spherical half hemisphere about 4 kilometers smaller in radius than the average.

The high gravity measures over the maria led the discoverers W. L. Sjogren and G. L. Muller to coin the term *mascon* for such a concentration of mass. Material of higher than average density clearly must exist in the depressed mascons for them to produce positive gravity anomalies. In fact, the near side of the Moon with the maria must contain higher density material near the surface than the mountainous far side. Furthermore, the Moon cannot be in a state of isostasy, where the total mass times gravity in small areas adds up to a constant gravity overhead. Otherwise there would be no gravity anomalies, no mascons. Conclusion: The Moon has considerable internal strength to support the increased mass under the maria and mascons. But more importantly, this structural strength must already have been attained when the maria were formed. The heavy basaltic lava must have flowed to fill the huge craters of the colossal impacts after the outer few hundred kilometers of the Moon were fairly cool and rigid. We shall see later how the lunar sample returns have supported and dated these postulated events.

Armed with this broader picture of lunar processes let us now return to a recurring controversy. Were *any* of the rills on the Moon made by water flow? We have seen that the simple geological processes of tension, settling, and cracking account for most. Some rills are lunar counterparts of terrestrial grabens (Fig. 92), where the cracks have been filled, resulting in a nearly level bottom. Several rills, however, give the strong impression of a flow pattern on a down slope. The most famous is named for the great German selenographer J. H. Schröter (1745–1816). The upper regions can be seen in Fig. 114, where the depth is some 1300 meters in the "cobra head." Others ap-

Fig. 114. A 74-km wide view of Schröter's Valley. The head is barely the crater's width from Aristarchus and extends for 350 kilometers. (Courtesy of the National Aeronautics and Space Administration.)

Fig. 115. Sinuous rills in the Harbinger Mountain plateau between Mare Imbrium and Oceanus Procellarum. Width of picture: 55 kilometers. (Courtesy of the National Aeronautics and Space Administration.)

pear in Fig. 115, one cutting across a range of hills. These rills and a few other unusual ones with sketchy tributaries appear to be explicable by combination of processes involving tension cracks, lava flows, and in some cases collapsed lava tubes. The fact that geologists are satisfied with such explanations and the fact that no lunar rocks show any evidence of ever having been exposed to water should end the controversy.

The greatest surprise of lunar exploration came from the seismographs planted by the astronauts for Apollos 12, 14, 15, and 16 (see Fig. 84 for locations). As we noted earlier, earthquake waves enable us to study the interior of the Earth much as x-rays allow us to study the interiors of human bodies. The Moon is unbelievably quiet seismically, having no winds, weather, waves, ocean tides, or humans to disturb its surface. As a result, the lunar seismographs can be made much much more sensitive than is practical on Earth. The smallest earthquakes correspond to the largest moonquakes.

As expected, lunar quakes were registered from rock slides, mostly on the mountain edges around the maria, a few probably from beneath the surface and occasionally from meteoroid falls. The surprise came from moonquakes that repeat themselves, usually to the finest detail in hundreds of oscillations. Furthermore, these repeating moonquakes are bunched in the lunar month, either when the Earth–Moon distance is minimum, at perigee, or else maximum, at apogee. As a third peculiarity the seismic triangulation proves that the sources or *epicenters* are located deep below the Moon's surface. More than two dozen have been pinpointed precisely by combining a number of records for each. The depths range from 700 to 1100 kilometers. They are observed only on the near hemisphere of the Moon, because the quakes are very weak and the waves appear to be dissipated in the central hotter regions of the Moon (see Fig. 116).

Apparently the strain of the Earth's tidal distortion on the Moon sets off these moonquakes at the times when this distortion is a maximum, at perigee. Correspondingly, the maximum tidal strain at apogee is about equivalent. Here we have the first direct evidence of tidal strains beyond the Earth, the magnificent example being volcano production on Jupiter's satellite Io by internal heating from great Jupiter's distorting power.

The deep moonquakes tell the story of seismic velocities and

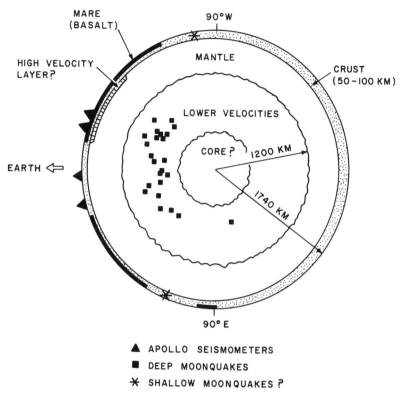

▲ APOLLO SEISMOMETERS
■ DEEP MOONQUAKES
✳ SHALLOW MOONQUAKES ?

Fig. 116. Depths of deep moonquakes and internal structure of the Moon. (Drawing by Joseph F. Singarella.)

the Moon's internal structure down to about two-thirds of its radius, confined to the near hemisphere. Models of the Moon's interior vary with the investigator but generally the velocity of the pressure waves is very slow in the upper crust, approximately 100 m/sec, and increases rapidly to about 7 km/sec at the bottom of the crust, some 50 to 100 kilometers deep. The velocity then increases to Earth-like values of around 8 km/sec in the mantle and probably decreases a bit below a depth of 500 kilometers. The shear waves are correspondingly slower by about 40 percent.

Below about 1100 kilometers the shear waves are mostly damped while the pressure waves are not much affected, indicating perhaps a partially melted central region. Very few moonquakes on the far side are strong enough to clarify the deep internal structure.

For the Earth the varying magnetic field proves that some circulation pattern exists in a fluid core or shell. The Moon shows no such general magnetic field, but certain regions of a few hundred kilometers in dimension have magnetic fields above a millionth of the average Earth field. The lack of general magnetism, the reaction of the Moon to the solar wind, the general internal temperature structure (little known), the rigidity to depths of about 1100 kilometers, density evidence from the crust, and gravity measures combine to indicate that if the Moon has a core, probably mostly iron, such a core is very small, perhaps less than 300 to 400 kilometers in radius. The central temperature may be the order of 1600°C.

The *crème de la crème* of the lunar space effort has been the return of actual lunar rocks and dust and their subsequent analysis with almost unbelievably subtle techniques. The six Apollo landings netted 382 kilograms of lunar material and the unmanned U.S.S.R. Luna 16 and Luna 20, some 130 grams, all from eight regions. The results of this truly magnificent effort divide broadly into three areas: (a) composition and structure relating to lunar chemistry and "geology," (b) isotope and age determinations relating to lunar chronology and evolution, and (c) solar wind implantation, micrometeorite cratering, magnetism, and other specialty studies relating to the lunar environment and history. No attempt can be made in this book to summarize the more than thirty thousand pages of scientific papers already published on these studies. A few paragraphs are devoted here to the chronology and compositional results while the structural and evolutionary aspects are largely condensed in Chapters 14 and 15.

A few of the radioactive isotopes most frequently used for rock age determinations are listed in Table 2, along with their atomic number, the most interesting or most useful stable decay (daughter) atoms, and the half-lives for decay. Intermediate short-lived radioactive atoms are omitted. Several types of ages can be determined. If the decay product is a gas such as helium, the measurement of the parent abundance, say uranium and thorium, determines the age since the rock was cool and solid enough to retain helium. The rubidium-strontium method is difficult to describe but depends on extremely precise measurement of several samples to give the age since the rock

TABLE 2. *A few radioactive isotopes.*

Atom	Products	Half life (millions of years)
Uranium 238	Lead 206 + He4	4,510
Uranium 235	Lead 207 + He4	713
Thorium 232	Lead 208 + He4	13,900
Iodine 129	Xenon 129	17
Rubidium 87	Strontium 87	47,000
Potassium 40	Argon 40 (11%)	1,300
Aluminum 26	Magnesium 26	0.74

was formed, not when it solidified. Similarly, the different isotopes of lead from many samples can be used to determine the age of the entire body. For the Earth and for meteorites as a group, lead isotope measures give the well-known age of 4.6 thousand million years or 4.6 aeons.

Lunar rock ages from the various missions then serve to calibrate other less direct methods, such as impact crater counts. These are very useful for dating areas of the Moon from pictures, paricularly large crater floors and maria lava flows. For microscopic scale dating, micrometeoroid craters and solar flare implantation methods are valuable. The high-energy solar ions damage the crystal lattice structure near the surface of a rock, the damage increasing with exposure, thus providing an exposure age measurement.

Most everyone was pleased to learn that the Moon has the same age as the Earth and meteorites, the accuracy being about a hundred million years, 0.1 aeon, or some 2 percent. The chronology of the Moon has been transformed from speculation to a fairly exact science, the uncertainty in time being this 2 percent or 0.1 aeon. Unfortunately, the accumulation time was also less than 0.1 aeon, so that little can yet be said with confidence about where and exactly how this great mass of material came together. In any case the outermost few hundred kilometers became a molten ocean. The lighter more refractory materials rose to the top, forming a thin crust. The lunar highlands are the result, composed of mineral types that are less frequent on Earth, *anorthites,* rich in the lighter elements calcium and aluminum combined with the ever-present silicon and oxygen.

The accumulation process by bombardment of infalling masses continued with decreasing intensity as the crust developed, puncturing it with great splashing craters that leave no record today. This colossal bombardment continued for about 0.6 aeons so that few, if any, present-day lunar features predate 4.0 to 4.2 aeons ago. The eastern maria such as Crisium, Tranquilitatis, Fecunditatis, and Serenitatis predate Imbrium, Procellarum, and Nubium, all representing great scars left by the late bombardment as it slackened. By 3.3 aeons ago the cratering impact rate had fallen to about its present value.

Radioactivity, far more intense then than now, was adequate to partially melt the interior of the Moon, even if it were relatively cool and solid 4.6 aeons ago. Radioactivity must have maintained or increased the molten layer under the thin crust, until the activity had fallen away and radiation cooling thickened the crust. Thus the heavier basaltic lavas, consisting of more iron and magnesium than the highland crust, flowed from beneath and slowly filled the maria basins. This filling was episodic, continuing irregularly until about 4.0 to 3.4 aeons ago for the older eastern maria and probably dying away completely on the Moon by 3.0 to 3.2 aeons ago.

Part of the large boulder sampled by the Apollo 17 astronauts goes back to 4.36 aeons but most of it was formed about 4.07 aeons ago, this on the edge of Mare Serenitatis. The rocks there average 3.90 to 3.98 aeons in age. On the west in Oceanus Procellarum the youngest basalts returned by Apollo 12 date to about 3.2 aeons.

By about 4.2 to 4.0 aeons ago the outer crust of the Moon had cooled enough to support large surface irregularities introduced by the great maria impacts. The lava layers and pools became smaller, to depths of 200 to 300 kilometers, after considerable separation of minerals by density and temperature. The later lava flows show this effect in denser basalts. As the maria basins filled, the outer mantle rocks supported these heavy mascons and did not further raise the nearby mountainous regions from beneath which the heavy lavas had flowed. Thus the isostatic maladjustment has remained permanently frozen in the Moon. The rills around the edges of the maria testify to small partial readjustments, that is, some sinking of the maria basins, but much less than complete isostasy would demand.

The cracking of the lunar surface—for example, the radial cracking from Mare Imbrium mentioned earlier—indicates that the interior of the Moon was heated after its original formation. The outer crust of the Moon was clearly in tension at the time of the Mare Imbrium event. A cooling Moon would have contracted generally to produce the "wrinkled prune" effect, so evident in the pictures of Mercury's surface. Thus the Moon was not formed as a molten ball.

For about 3 aeons the Moon has looked essentially as it does today, except for the addition of a number of individual craters, particularly the ray craters. Earlier ray craters were surely present, but the rays have been erased from visibility. The Copernicus crater is thought to have formed some 0.8 aeons ago and Tycho perhaps as recently as 0.11 aeons, only 110 million years ago.

Tide-raising forces and the evidence presented in Chapter 7 for the lengthening of the Earth's day indicate that the Moon formed much closer to us than it is today. Sir George Darwin (1845–1912), son of the great naturalist, theorized that the Moon was once in contact with the Earth, the two bodies having a period of rotation of about 4 hours, and that subsequently they separated because of tidal friction. Darwin's complete theory cannot be supported because viscosity in the Earth-Moon would have prevented their separation. There is, however, no clear indication concerning the actual distance from the Earth at which the Moon was formed. Probably it formed at much less than half its present distance and possibly quite close to the Earth. Once there was hope that a fossil tide might be preserved in the Moon but the Moon was not rigid enough when near the Earth.

Because tide-raising effects vary as the inverse cube of the distance, the rate of the Moon's travel from the Earth was very much faster in the early stages than it is today. Indeed, there may have been a time when the tide-raising effects on the Earth were far greater than those that we now observe. Kuiper even suggests that these tides were so high—of the order of a kilometer or two—that they had profound effects on the topography of the ocean bottoms, assuming that the Earth indeed had water at this early stage in its development.

In Chapter 15 we shall speculate a bit more on possible mechanisms of lunar formation and evolution. Direct exploration of the Moon is continually answering old questions and raising many new ones. Landings on the Moon have already penetrated some of the secrets of the Goddess of the Night. Her ancient skin carries a record that predates any now left on Earth and her story reaches back to the days when the Earth was new.

10

Pluto, Mercury, and Venus

Pluto

The most distant planet yet discovered appears to be a dwarf interloper among the giants. Pluto, a planet that is probably smaller than the Moon, appears to be rotating with a period of 6.39 days, according to photometer observations by J.-S. Neff, W. A. Lane, and J. D. Fix. The light variation is about 25 percent.

Pluto's gravitational mass as calculated has rapidly decreased over the years. Improved and continuing observations of Uranus and Neptune reduce the evidence for any perturbations on these planets by Pluto (see Chapter 3). The effects of small systematic errors in the early observations of Uranus and Neptune are being reduced as the observational time span expands. Hence the prediction of Pluto's existence was made purely by chance. Kuiper calculated Pluto's apparent diameter to be the order of 0.2–0.3 arcseconds—really a bit beyond the resolving power of his method—and its corresponding diameter to be about 6000 kilometers, 1.6 times that of the Moon. Ian Halliday of Canada and other observers set a similar upper limit by failure to see any occultation of a star 0.125 arcseconds from Pluto.

J. W. Christy's discovery of a probable satellite of Pluto, tentatively called Charon, from elongated images of the planet, provides more realistic estimates of Pluto's mass and diameter. The satellite's period of 6.39 days, equal to the rotation period of the planet, and an estimated mean distance from Pluto of 20,000 kilometers (0.7 arcseconds at Pluto's mean distance) leads to a mass of 0.21 of the Moon's mass for Pluto. Hence if the planet has a density of 3.34 times the density of water, equal to that of the Moon, its diameter becomes 0.6 of the Moon's diameter, or 2100 kilometers. We may take this as a lower limit, as it corresponds to an albedo of 1.0, making Pluto a perfect reflector. Correspondingly, if Pluto is made from the aggregation of comets—mixtures of ices and dust—then its density might be in the range 1.0 to 1.8 times that of water, and its diameter would be 3200 to 2600 kilometers, for albedo of 0.4 and 0.6 respectively. Polarization measures by L. A. Kelsey and J. D. Fix suggest an albedo below 0.25, supporting a low density for Pluto. On the other hand, D. P. Cruikshank and colleagues find infrared bands in the spectrum of Pluto that indicate the presence of methane (CH_4) frost on the surface. At an expected temperature of some −228°C on Pluto, methane could well freeze out. A fairly high albedo might result. All of this is consistent with Charon and its orbit being real and with Pluto being of density 1.0 to 1.8 times water and of diameter some 3000 kilometers, actually smaller than the Moon. Pluto is certainly an arid, frigid, and inhospitable little world.

If Charon is physically like Pluto and is observed to be six times fainter, it must have about $\frac{2}{5}$ Pluto's diameter and about $\frac{1}{15}$ of Pluto's mass. It appears to move in a highly inclined orbit and it revolves with the rotation period of Pluto. If the planes of these two motions are the same, Charon may hover over a spot on Pluto's equator, a unique but stable configuration in the solar system.

The suggestion has been made that Pluto is not truly a planet at all but a lost satellite from Neptune. This question cannot be answered until we know more about the mechanisms whereby satellites develop about planets. At least we need not worry

about Pluto's destruction by collision with Neptune, even though their orbits overlap. E. Öpik, C. J. Cohen, and E. C. Hubbard have shown that Pluto and Neptune are held gravitationally in a repeating orbital cycle of nearly 20,000 years, so that they can never collide. If this relation holds rigorously, it would suggest that Pluto may not have once been a satellite of Neptune. It may be a relative of the unique object, Chiron, which moves in an orbit between Saturn and Uranus. Chiron, perhaps some 200 kilometers in diameter, may be either a comet or an asteroid, truly a mystery object. Its orbit is unstable over tens of thousands of years so that we have no idea as to its past history nor, indeed, its future. Undoubtedly there are other such small bodies in the outer solar system, mavericks surviving by chance for 4.6 aeons.

Mercury

Mercury is the fourth brightest planet, at its best nearly equaling Sirius in brilliancy and being exceeded only by Venus, Mars, and Jupiter. Nevertheless, Mercury is a very difficult object to observe because of its small orbit and concomitant proximity to the Sun; the greatest possible *elongation* (apparent angle from the Sun; see Appendix 2) is 28°. At this most favorable position the phase corresponds to the quarter moon; the full phase occurs at superior conjunction when Mercury lies beyond the Sun, nearly in line with it. After sunset or before sunrise Mercury is always low in the sky, a situation that limits night observations to a short interval. In addition, the turbulence of our atmosphere at low altitudes produces poor "seeing." Hence Mercury, to a great extent, is observed in full daylight, scattered sunlight being eliminated as much as possible by suitable screens. The NASA Mariner 10 flybys in 1974 sent back superb pictures covering half the surface of Mercury, really our first "solid" information about Mercury's surface (Figs. 3, 117, 118).

Visual observers long ago agreed that Mercury keeps the same face toward the Sun and thus rotates on its axis in its pe-

Fig. 117. Photomosaic of Mercury by Mariner 10, centered near the equator and reaching to longitude 190° on the left beyond which the Caloris basin is centered. (Courtesy of the National Aeronautics and Space Administration.)

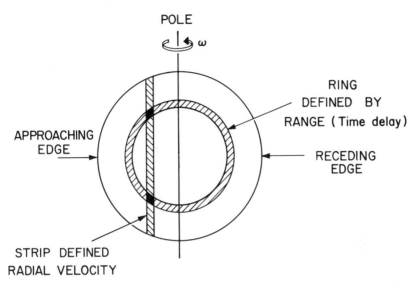

POLE

RING
DEFINED BY
RANGE (Time delay)

APPROACHING
EDGE

RECEDING
EDGE

STRIP DEFINED
RADIAL VELOCITY

Fig. 119. A spherical planet illuminated by radar. The radar beam, broader than the apparent planetary diameter, is limited by time delay to a circle on the apparent disk and by frequency to a strip parallel to the rotation axis of the planet.

riod of revolution, 88 days. Radar observations from the great 1000-foot dish at Arecibo, Puerto Rico, dispelled this long-standing illusion in 1965. Radar waves, scattering off a rough sphere like Mercury, can measure the range to a ring (Fig. 119) on the surface by control of the time lag from pulse transmission to pulse reception (10^{-6} second error equals 150 m). The diameter of a planet or satellite can be measured quite precisely by the length of the pulse echo corrected for the inherent length of the transmitted pulse. In addition the frequency or wavelength of the returning echo will be changed by the relative velocity between the source and target, the Doppler-Fizeau effect, described on p. 257. Thus, a rotating planet presents a higher returning frequency on its approaching side than on the receding side. A chosen frequency band can then be selected to measure the echo from a strip parallel to the rotation axis on

Fig. 118. Photomosaic of Mercury by Mariner 10. This picture overlaps Fig. 117 on the left but common features are difficult to match because of the curvature. About one half of the planet is covered in these two pictures. (Courtesy of the National Aeronautics and Space Administration.)

the apparent planetary disk. By choices of both time delay and received frequency, the combined echo from two areas (black in Fig. 119) common to the ring and strip can be measured. If the radar dish is large enough at a given frequency or, if not, by interference between two receiving systems, the beam can be narrowed enough to distinguish between the two dark areas. Thus radar maps of the Moon now rival the best Earth-based photographs.

For Mercury the radar-Doppler measures showed that the sidereal period of rotation is only 59 days direct, not 88 days as long believed. Immediately it became apparent that the period was nearly two-thirds the period of revolution and that the angular velocity about the Sun at perihelion would be very close to the average angular rotation. Since tidal forces vary as the inverse cube of the distance, the solar gravitational control on a slightly elongated Mercury body would mostly take place near perihelion and could stabilize the rotation period at two-thirds of the period of revolution. If Mercury's rotation had originally been much more rapid and had been reduced by solar tidal friction, it could finally have "locked-on" when it reached a period near 59 days. The best measured periods by radar, by the reinterpreted telescopic observations, and by Mariner 10 confirm the precise two-thirds resonance period of 58.6457 days.

To the optical astronomers who had developed maps of Mercury consistent with a period of 88 days, the radar result came as a profound shock. The maps, however, could be quickly revised on the basis of the new period. Because of the two-thirds relation, the maps were actually correct half of the time, poor seeing being blamed for the inconsistencies. The optical markings actually show little correlation with the geologic features.

The consequence of the $\frac{3}{2}$ resonance is to give Mercury an extraordinary long day of 176 Earth days. While Mercury rotates completely with respect to the stars in about 59 days it moves $\frac{2}{3}$ of the distance around its orbit. This is $1 - \frac{2}{3}$ or $\frac{1}{3}$ of its day, making the total day 3 times its sidereal rotation period and twice its period of revolution. In Fig. 120 we look down from the pole of the orbit with the body of Mercury fixed at the center of the diagram. Choosing the arbitrary zero of longitude at perihelion noon, we see how the Sun appears to move around Mercury in the outer curve. At perihelion Mercury turns slightly slower

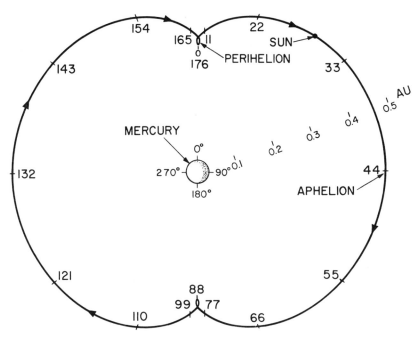

Fig. 120. Mercury's day lasts for 176 Earth days. Projection of solar distance on Mercury's equator with arbitrary longitude marks.

than it swings around the Sun, so that the Sun moves a little eastward, as seen from Mercury's equator, before it continues its westward daily motion. At longitudes of exactly 90° and 270° on the equator in Fig. 120 the Sun peeks above the horizon for an Earth day or two, then sets and rises again to remain ablaze again for nearly a Mercurian year; it then proceeds to set twice. The Sun appears about 50 percent larger in diameter at perihelion than aphelion and delivers more than twice the heat to Mercury's surface. The pole of rotation, measured by radar, telescope, and Mariner 10, is perpendicular to the orbit within the measuring accuracy of 2 to 3°.

The dearth of atmosphere, expected from the low surface gravity and low velocity of escape (4.2 km/sec), coupled with high temperature on the day side and lack of twilight extension of the horns in the crescent phase, is confirmed by Mariner 10 direct spectroscopy and by electron density measures from radio occultation. Helium and hydrogen atoms are present but the surface pressure is less than a million-millionth of the

Earth's at the subsolar point on Mercury. On the cold night side, the pressure may be an order or so of magnitude higher, but still trivial in amount. If Mercury is exuding gas from its interior, the rate is believed to be less than a gram per second.

The temperatures on Mercury are about what we might expect for the Moon, were it moved into Mercury's orbit close to the Sun at a distance averaging only 0.39 AU. The immensely long nights drop the temperature to a minimum of about −173°C or lower while the correspondingly long days produce a maximum of some +430°C near the subsolar point. A typical night–day variation is thus some 600°C! Tin, lead, and even zinc melt at 430°C. Hence Mercury rather than Pluto might well have been named after the god of Hades.

Mercury's diameter is 4878.0 kilometers, now extremely well determined by radar and by Mariner 10 radio occultations. This is only 1.40 the diameter of the Moon, but Mercury's mass is 4.5 times that of the Moon, leading to a mean density of 5.43 ± 1 times water, almost that of the Earth, 5.52. But Mercury is little compressed, much less than the Earth, making Mercury the intrinsically densest planet. It must contain an unusually large fraction of iron (density 8.8 water), the most abundant of the denser elements. Indeed the fraction of iron by mass must be some 60 percent or more.

Describing the surface features of Mercury would be almost repetitious of the Moon account. Comparison of Mariner 10 pictures with lunar pictures shows the basic similarity. This similarity extends to the albedo, to scattering properties of the surface, and even to the thermal properties of conductivity and thermal inertia or heat retention. We may conclude that Mercury is covered with fine soil, compacting with depth, as should occur on an airless body whose upper layers have been gardened by impact cratering for some three or four aeons.

The differences between Mercury and the Moon, though small, are of some interest. The mountains on Mercury are not as high as those on the Moon. The Mercurian maria-like basins are surrounded with lightly cratered plains, but the albedo differences are less than between the lunar maria and the highlands. On the other hand, the rays from more recent craters, never visible from Earth, are even more contrasty than for the Moon. The rich iron content of Mercury is suspected of showing on the surface as iron-titanium-rich glass, reducing the compositional differences between highly cratered (highland-like) areas relative to the maria. The rays, rough and broken, are more comparable to lunar rays.

Many of the larger craters on Mercury now bear the names of famous authors, artists, and musicians such as Homer, Shakespeare, Tolstoy, Rodin, Titian, Renoir, Bach. Beethoven's is the largest, of diameter 625 kilometers.

The Caloris basin in Mercury (Fig. 121) is the counterpart of the great Imbrium basin on the Moon. Some 1300 kilometers in diameter, Caloris presents central regions that are smooth plains but highly cracked and ridged. On the edge are irregular mountains, rising up to 2 kilometers in height. Radial valleys and ridges extend out another 1300 kilometers, much as around Imbrium. Evidence of a basic ejecta blanket extends much farther. The shock of this impact was so great as to leave a peculiar surface pattern at the antipodal point (Fig. 122). Probably the seismic shock wave focused there to produce the degraded hills and linear valleys. A focusing of the crater ejecta at the antipode is possible but seems physically less likely.

Apparently the Caloris event produced other planetwide effects. J. E. Guest and Donald E. Gault conclude that all the young, sharp craters smaller than 30 kilometers in diameter were formed afterwards. Whether the older smaller craters were degraded by the Caloris event or by prior events and processes is not clear. Because no time scale is yet available for Mercury, we can only wonder whether the Caloris event was related in time with the great basin-forming impacts on the Moon. A large asteroidal-type body in the inner solar system might well have been broken by a collision to provide many large pieces for impact on the Moon and terrestrial planets. A number of planetary astronomers find this idea attractive.

The lack of relief on Mercury as compared to the Moon probably arises from its greater gravity (2.3 times) and perhaps because the major cratering period may have occurred while Mercury was more plastic. In any case Mercury craters differ systematically from the lunar craters in that the secondary cratering around the walls is more intense but more restricted in diameter. Also the occurrence of terraces on the walls and peaks on the floors of the larger craters is more frequent on

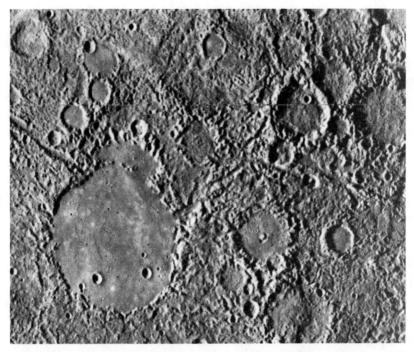

Fig. 122. The Crater Petrarch, of diameter 160 kilometers at longitude 27°, latitude 40° S, on the edge of the region antipodal to the Caloris Basin. The crater and valley can be seen on Fig. 118, center a little to the right and below. (Courtesy of the National Aeronautics and Space Administration.)

Mercury. The rims of the larger craters are relatively lower. All of these effects may be ascribed to the gravity differential. Mercury shows no evidence of mountain building or plate tectonics like that on Earth.

Mercury produces a bow wave in the solar wind at a distance of perhaps one radius from the surface of the planet. Thus Mercury appears to possess a small magnetic field less than one percent of the Earth's. Because of the slow rotation period, no large field would be expected even if Mercury does possess a sizable liquid core.

Mercury's orbit is next to Pluto's in terms of high inclination

Fig. 121. Photomosaic of half the Caloris Basin on Mercury by Mariner 10. Compare left of Fig. 117. (Courtesy of the National Aeronautics and Space Administration.)

to the plane of the ecliptic, 7.0°, and eccentricity, 0.21. At perihelion its orbital speed reaches 58 kilometers per second. As noted earlier, this rapid motion and high eccentricity have enabled Mercury to provide one of the three astronomical verifications of the Einstein theory of general relativity. The direction of Mercury's perihelion advances some 43 seconds per century more than is accounted for on the basis of planetary perturbations. Einstein's relativity theory, however, predicts the observed rate within its accuracy of measurement. Mercury is thus a major contributor to modern science.

Venus

Venus is both the "evening star" and the "morning star," the Hesperus and Phosphorus of antiquity. It is the most brilliant object in the sky, except for the Sun and Moon. Venus is often visible in the daylight and capable of casting shadows at night. Only 144 days elapse from the evening elongation, when Venus is the first object to be found in the evening twilight, until the morning elongation, when it is the last "star" to disappear in the Sun's morning glow, while 440 days are required for Venus to revolve beyond the Sun and return again to its evening elongation (the geometry can be seen in Fig. 4). Its true period of revolution about the Sun is, of course, much shorter, only 224.70 days. At minimum distance it becomes our nearest planetary neighbor, some 42,000,000 kilometers distant. But then optical observations are difficult because it appears so close to the Sun.

Venus is truly the Earth's sister planet, nearly of the same size and mass. The magnification of even a small telescope suffices to resolve the brilliant point of light into a silvery disk, somewhat diffuse at the edges because of the unsteadiness of our atmosphere, but showing the crescent phases like the Moon. When the crescent is thin the horns appear to extend more than half around the disk, as though the irradiation of the brilliant surface were producing an optical illusion. In the extreme situation, however, when Venus lies nearly in the line between us and the Sun, a faint circle of light can be seen entirely around the disk. This twilight arc is shown in the photographs of Fig. 123. A deep atmosphere on Venus deflects the sunlight around the edges of the disk by refraction and scattering.

Fig. 123. Twilight arc around Venus. Note the extension of the crescent completely around the disk in the left-hand photograph and the bright extension at the top in the right-hand photograph. (By E. C. Slipher, Lowell Observatory.)

But why do we not see clouds in the atmosphere or else surface markings on the globe itself? Under the best observing conditions, when our atmosphere is clear and steady, only the haziest suggestions of markings can be seen by the most expert observers—"large dusky spots," as Barnard called them. These faint patches, too indefinite to be drawn well, are impermanent. Photographs in the long wavelengths of infrared light have also proved unsuccessful in registering details, not withstanding the haze-penetrating power of the infrared light. It was not until F. E. Ross (1874–1960) experimented with the other extreme of the color spectrum, ultraviolet light, that details on Venus could be photographed. To everyone's surprise—because ultraviolet light is markedly useless for clouds on the Earth—Ross was successful in registering great hazy cloudlike formations in the atmosphere of Venus that change from day to day. Figure 124 illustrates the complex structure of Venusian clouds as imaged also with near ultraviolet light by the NASA Mariner 10, in February 1974. The cloud pattern changes drastically with time and position on the planet in the many images that have been made from space near Venus. A four-day rotation observed in the Venus ultraviolet clouds was once taken to repre-

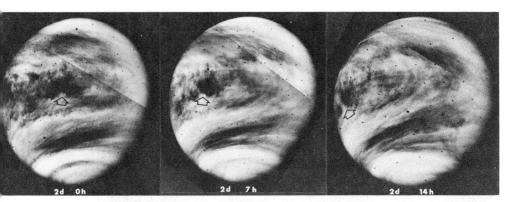

Fig. 124. Venus cloud imaged in ultraviolet light by Mariner 10. The time span of 14 hours shows the 4-day rotation of the changing cloud pattern. (Courtesy of Michael J. S. Belton and the National Aeronautics and Space Administration.)

sent the rotation of the planet itself. It is now believed to be the result of the complicated circulation pattern in the atmosphere.

The remarkable feat of radar to Venus by the Millstone Hill radar of the Massachusetts Institute of Technology (see Fig. 34), the Goldstone facility at the California Institute of Technology, and a radar in the Soviet Union first clarified the situation. On page 175 with Fig. 119 we discussed radar techniques for measuring planetary surface features, dimensions, and rotations. For Venus the result was at first difficult to accept. The rotation is *retrograde* with a sidereal period of 243.0 ± 0.1 days, the pole lying within 3° of the orbital pole. A period of 243.16 days would make three rotation periods of Venus resonate with two revolutions of the Earth about the Sun, as though Earth tidal forces might have "locked-on" to the body of Venus. Unlikely as the idea may seem because of the small forces involved, we cannot regard the "locking-on" as impossible. In any case, the slow retrograde rotation is firmly established and theories of planetary evolution must henceforth allow or account for this anomaly in planetary motion. The daily cycle on Venus is 117 Earth days.

Radar shows that the surface of Venus reflects about twice as well as that of the Moon or Mercury, 0.12 of perfect reflection averaged over a wide range of radio frequencies. The surface is significantly smoother, however, than that of the Moon or Mer-

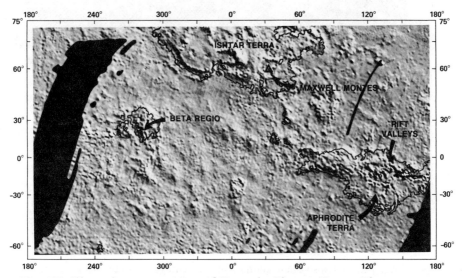

Fig. 125. The radar contour map of Venus by Pioneer Venus. The map projection exaggerates the areas away from the equator but preserves the shapes of smaller features. The black areas represent gaps in the coverage. The names of the features may be subject to change by the International Astronomical Union. (Courtesy of the National Aeronautics and Space Administration.)

cury; parts of it are very smooth, although the general scattering law (delay curve) is similar to that of the Moon.

NASA's Pioneer Venus, having a radar in orbit about Venus, at last penetrated the perpetual clouds in 1980 to give our first real view of Earth's sister planet. In Fig. 125 we see the radar map, covering about eighty percent of the surface, and in Figs. 126 to 128, an artist's conception of the most conspicuous features, based on the radar measures. The highland areas are like continents on Earth while the lowlands, corresponding to our ocean basins, cover only a sixth of the surface, compared with two-thirds on Earth. The Beta Regio formation (Fig. 126) appears to be two huge shield volcanos, dwarfing our Hawaiian counterparts in area, but not in height; the Beta volcanos rise about four kilometers from the surrounding plains.

The two "continents" on Venus, Aphrodite Terra and Ishtar Terra (Fig. 127), are comparable to the continental United States in size; most of the planet consists of an irregular rolling plain possibly spotted with some sizeable crater-like formations and what may be other volcanic mountains much smaller than Beta

Fig. 126. Artist's visualization of the Beta Regio shield volcanos on Venus. Two of the Russian Venera spacecraft landed directly to the east of Beta and found the rocks to be basaltic. (Courtesy of the National Aeronautics and Space Administration.)

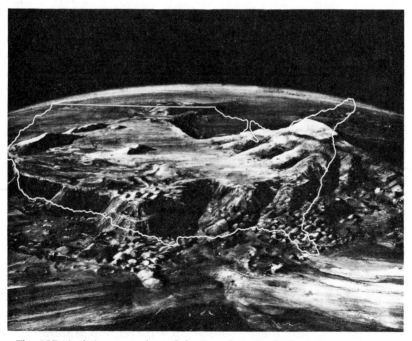

Fig. 127. Artist's conception of the "continent" of Ishtar Terra on Venus, with the outlines of continental United States superimposed. The projection is distorted at North. (Courtesy of the National Aeronautics and Space Administration.)

Fig. 128. Artist's conception of the great Rift Valley at the eastern end of Aphrodite Terra on Venus. The slopes are highly exaggerated, the maximum being about seven degrees, which is steep for Venus. The Rift is 280 kilometers wide but only about 2 kilometers deep. (Courtesy of the National Aeronautics and Space Administration.)

Regio. The Ishtar Terra continent, although a bit smaller than Aphrodite Terra, stands out because of its mountain massif, the Maxwell feature, which rises nearly eleven kilometers above the mean height or "sea level"—more than two kilometers higher than Mt. Everest on Earth. The mountain region—the roughest on Venus—is conspicuous to Earth radars in spite of its high latitude on the planet, where only very rough surfaces can produce an echo. The Ishtar plateau to the west of Maxwell is twice as large as the Tibetan plateau and higher. It seems to be relatively smooth, possibly covered with "young" lavas.

Aphrodite Terra is unique for two great rift valleys on its eastern end, some twenty-two hundred kilometers in length (Fig. 128). The rifts are below "sea level," like our midoceanic rifts, and very much like the great Valles Marineris, which we will meet on Mars. Extremely rough mountain ranges define the northeast and northwest boundaries of Aphrodite Terra. The western mountains are the highest, rising a respectable seven kilometers above the plains near them and nearly eight kilometers above "sea level."

We see that the topography of Venus has many earthlike features, indicative of a geologically active planet. We have yet to find evidence of current volcanic activity, if any, and also to establish the frequencies of meteoritic craters to give us possibly a dating system.

Radio techniques provided an even greater surprise about Venus than its surface features and its retrograde rotation. Venus is very hot! Infrared measures of temperature at the cloud tops have been consistently cool, near −38°C both on the sunlit and on the dark sides. This value corresponds well to that at the tops of high Earth clouds. But the radio thermal measures are remarkable. At the very shortest wavelengths, 3 millimeters, the temperature has begun to rise above the infrared value. At a wavelength of 2 centimeters and at longer wavelengths the Venusian temperature has reached the amazing value of more than 310°C. This temperature was confirmed by the U.S.S.R Venera 4 spacecraft that actually entered the atmosphere at 11 kilometers per second and released an instrumented probe on a parachute. The maximum pressure also seemed astonishing at the time, 12 to 22 atmospheres. Other Veneras, NASA Mariners, NASA Pioneers, and especially, in 1978, the Pioneer Venus orbiter and microprobe missions have led to even more elevated values (Fig. 129). Evidently the great heat incapacitated the first probe before it landed.

The surface temperature of Venus is 457°C with an uncertainty of perhaps 20°C; it varies a bit from night to day and falls very rapidly with altitude. A number of the normally solid elements, including cadmium, lead, tin, and zinc, would be molten on Venus while sulfur would boil except for the pressure, which keeps it and sulfuric acid in the liquid form. For comparison, ordinary kitchen baking temperatures rarely exceed 260°C. The pressure on Venus is phenomenal by Earth standards, some 90 atmospheres! It corresponds to the pressure in the ocean at a depth below 900 meters, well beyond the reach of present-day scuba divers.

Not only the amount but also the composition of the Venus atmosphere is surprising. The first evidence for this fact came

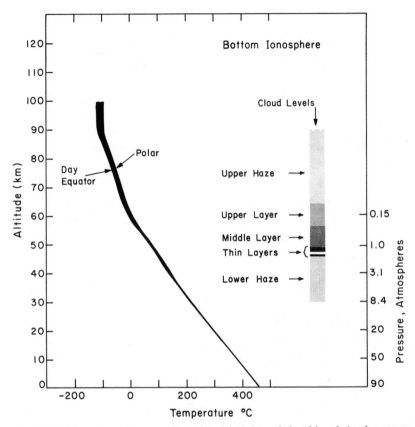

Fig. 129. Temperatures, pressures, and heights of cloud levels in the atmosphere of Venus. (By Joseph F. Singarella.)

from an Earth-based spectrograph, an instrument that separates light into its fundamental colors and photographs the entire sequence, from the ultraviolet through the blue, green, yellow, red, and infrared. The light to be analyzed first passes through the entrance slit of the spectrograph (Fig. 130) then through a lens to a prism, the heart of the instrument. The prism or grating disperses the light into its constituent colors, forming a spectrum. The spectrum is identical in character to a rainbow except that the colors are much better separated. A second lens of the spectrograph serves to focus the spectrum on the photographic plate, or sensitive imaging device. With a long exposure the spectograph registers light much too faint to be seen with the naked eye.

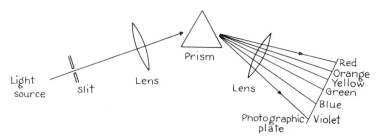

Fig. 130. Schematic drawing of a spectrograph. The optical functions of the lenses are not indicated.

The spectrograph, which is now extended in principle to the infrared and radio regions, solves a number of our difficult problems, one of the most important being the identification of chemicals in a gaseous mixture. The light that escapes from a luminous or absorbing gas reveals the character of the gas. Each atom or molecule is a vibrating pulsing entity; the vibrations are those of the elementary negatively charged particles, the electrons, which are held in miniature orbits or energy states about the heavier positively charged nuclei of the atoms. Two or more atoms in a molecule also rotate and vibrate around each other, while the electrons in the atoms spin about in their complex gyrations.

All of these motions and vibrations in atoms and molecules represent discrete amounts of energy, exceedingly minute but characteristic for each kind of atom or molecule. When an atom loses energy by a change in the rate of vibration, the energy is radiated into space as a *photon* of light (or *quantum,* the smallest unit of radiant energy), with a definite amount of energy, a certain color to the eye (if visible), and a certain wavelength of vibration. These waves of light belong to the same family as radio waves but are much shorter. The waves of red light are about $\frac{1}{1600}$ millimeters or 0.6 microns long, while those of blue-green light are shorter. Infrared or heat waves may be several times longer; ultraviolet light waves are about half this length or shorter.

The compressed incandescent gas near the surface of the Sun sends out all colors and therefore all wavelengths of light. When such continuous light traverses a layer of cool gas, such as our atmosphere or the atmosphere of Venus, the vibrating atoms and molecules are activated and steal from the beam precisely those wavelengths characteristic of their rates of vibration. When we analyze the light with a spectrograph we measure the wavelengths that are missing, and so identify the atomic or molecular thieves who are responsible for the loss. The missing wavelengths appear as dark lines in a spectrum.

Earth-based spectrographs repeatedly failed to find evidence for water vapor on Venus or to show identifiable lines and bands. Walter S. Adams (1876–1956) and Theodore Dunham at the Mount Wilson Observatory discovered new infrared absorptions that were unknown from laboratory studies. To check a theory that the unknown absorptions might arise from ordinary carbon dioxide Dunham filled a 60-foot pipe with carbon dioxide compressed to ten times the pressure of our atmosphere. When artificial light was sent down the tube and reflected back to the same spectrograph used for the Venus spectra, identical absorptions were obtained. A spectrogram of Venus is compared with the solar spectrum in Fig. 131.

Space missions to Venus have confirmed the enormous carbon dioxide content of the atmosphere. The Pioneer Venus sounder probe on December 9, 1978, provided the most complete data. Table 3 lists the composition by volume as obtained by the gas chromatograph of Pioneer Venus and by the mass spectrograph of Venera 10. Although nitrogen does not dominate Venus's atmosphere as it does the Earth's, the total content there may be three times larger or more. Water vapor, highly variable in the Earth's atmosphere, may exist, nearly comparable in absolute quantity. Oxygen is truly deficient, the total amount adding up to around 3 percent or less of the total in the

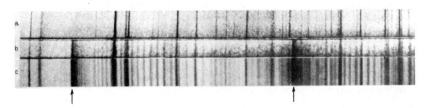

Fig. 131. Spectrograms (*a*) of the Sun; (*b*) of Venus; (*c*) Venus widened. Note the infrared absorptions of carbon dioxide (*arrows*), strong in the spectra of Venus but absent in the spectrum of the Sun. (Photographs by T. Dunham; courtesy of Yerkes Observatory.)

TABLE 3. *Atmospheric composition of Venus.*

Gas	Percent by volume	
	USA[a]	USSR[b]
Carbon dioxide	96.4	97
Nitrogen (N_2)	3.4	1.8
Water	0.002 to 0.52	0.1 to 1.0
Hydrogen, H_2	0.03	
Sulfur dioxide	0.018	~0.1
Carbon monoxide	0.003 to 0.007	—
Oxygen (O_2)	0.007?	—
Argon	~0.0065	—
Ammonia	—	0.01 to 0.1
Neon	0.0005 to ~0.0020	—
Helium	0.0001 to 0.001	

a. By V. I. Oyama, G. C. Carle, F. Woeller, and J. B. Pollack.

b. A. Surkov and M. Ya. Marov for source. From Earth-based spectroscopy, hydrochloric acid and hydrogen fluoride appear as traces. Helium and the hydrogen atom predominate in the very high atmosphere, especially at night.

Earth's atmosphere. Some exciting results concerning the isotopes of argon and neon will be discussed in Chapter 15 as they relate to the comparative evolution of Earth and Venus.

A high temperature at the bottom of an atmosphere containing a great deal of carbon dioxide is not a very surprising result. The gas is very transparent to all visual light, and, unlike oxygen, to ultraviolet light. Carbon dioxide, however, absorbs heat (far-infrared) radiation extremely well. The result is that the greenhouse effect should be powerful in heating the surface of Venus. Much of the Sun's energy can enter as visual light, while the radiation from the heated surface is trapped by the carbon dioxide. This identical process helps to regulate the surface temperature of the Earth. Carl Sagan has shown, however, that a little bit of water vapor can help "seal up the chinks" in the infrared spectrum of carbon dioxide through which heat might escape from the surface of Venus. The amount of water observed is adequate for this purpose and the trace of sulfur dioxide helps fill the remaining "chinks."

On Earth the water acts as a catalyst, enabling carbon dioxide to combine with silicate rocks, thus "fixing" the carbon dioxide in carbonate rocks. A layer some 200 meters deep has been formed around the Earth in geological times. This carbon plus that in the oceans, in the pores of the soil, and in the biosphere add up to about the amount of carbon dioxide now present in the atmosphere of Venus. Thus the surprise in the atmosphere of Venus is not the presence of carbon dioxide but the absence of primitive water.

Again, oxygen in our atmosphere comes from the water dissociated by solar ultraviolet light, the hydrogen being lost to space because of its low molecular weight. Thus the lack of water on Venus prevents the production of much oxygen, in agreement with the observations.

The composition of the clouds on Venus has remained a celestial challenge for many decades. At last the evidence is clear: the clouds are composed largely of concentrated sulfuric acid droplets! The high haze above 60 kilometers and the low haze below 48 kilometers consist largely of particles with diameters close to 1 micron, twice the wavelength of blue-green light (see Fig. 129). In the denser clouds between these levels the distribution of particle sizes is much wider, peaking in mass between 5 and 10 microns. Perhaps some solid sulfur particles are present. The daylight probe of the 1978 Pioneer Venus mission stirred up some dust when it landed, the dust settling out in a few minutes. What kind of dust is it? Lightning, comparable to terrestrial lightning, is observed on Venus. Does it occur because of sulfuric "rain" storms?

Although the clouds on Venus completely obscure our view of the surface except by radar, they are not as black as one might suppose. An appreciable amount of solar radiation must reach the surface to maintain the greenhouse effect. The U.S.S.R. Venera landers measured the surface brightness at about 10 percent of that on Earth, like that on an overcast day. About 6 percent of the solar flux reaches the surface. Thus it was possible for the Venera 9 and 10 landers to make the extraordinary pictures of the surface shown in Figs. 132 and 133 using only natural light.

Three Veneras also measured the natural radioactivity of the surface rocks. The activity was comparable to that of basaltic rocks for Veneras 9 and 10 but somewhat closer to that of granite for Venera 8. The density of the rocks at a depth of several dozen centimeters could be deduced by Venera 10, at 2.7 ± 0.1 water density, much smaller than the mean density of Venus (5.25). The value is typical of basaltic rocks.

The rocks in Figs. 132 and 133 show some evidence of ero-

Fig. 132. U.S.S.R. Venera 9 panoramic photograph of the surface of Venus, October 22, 1975 (computer enhanced). The boulders are roughly 50 centimeters in size. The field of view is 180° looking down at an angle of 50°.

sion or weathering. Wind scouring is certainly not conspicuous and might not be expected on the basis of the very light winds measured near the surface, 0.4 to 1.3 meters per second up to an altitude of several kilometers. From 10 to 50 kilometers altitude the wind rises to an average of 50 to 60 meters per second, typical of considerably higher altitudes. The nature of any chemical weathering of the rocks is not known. Possibly volcanic action and seismic disturbances are the major sources of rock degradation on Venus.

The NASA Mariner 2 (Fig. 134) was the first spacecraft to fly by Venus, at a distance of some 34,800 kilometers. One of its major results was to demonstrate that Venus has an almost negligible magnetic field, now found to be only a few times 10^{-5} that of the Earth. Hence the nose of the bow wave of the solar wind occurs at a distance of only about 1.5 radii from the center of Venus, about a tenth of the corresponding distance on Earth. The magnetosphere reaches little above the ionosphere and

Fig. 133. U.S.S.R. Venera 10 panoramic photograph of the surface of Venus, October 25, 1975 (as the transmission was received). In the center can be seen the gamma-ray densitometer on the probe, about 40 centimeters long. The rocks of the Venera 10 site appear to be more degraded than those of the Venera 9 site.

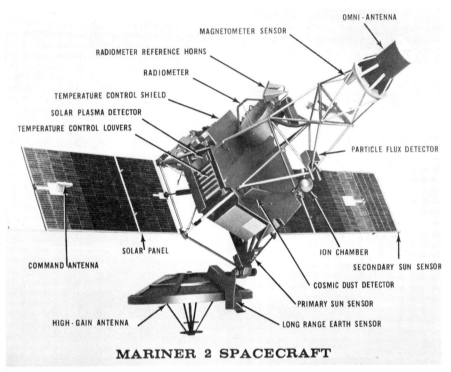

OMNI - ANTENNA

MAGNETOMETER SENSOR

RADIOMETER REFERENCE HORNS

RADIOMETER

TEMPERATURE CONTROL SHIELD

SOLAR PLASMA DETECTOR

TEMPERATURE CONTROL LOUVERS

PARTICLE FLUX DETECTOR

SOLAR PANEL

COMMAND ANTENNA

ION CHAMBER

SECONDARY SUN SENSOR

COSMIC DUST DETECTOR

PRIMARY SUN SENSOR

HIGH - GAIN ANTENNA

LONG RANGE EARTH SENSOR

MARINER 2 SPACECRAFT

Fig. 134. The Venus spacecraft, Mariner 2, that approached Venus on December 14, 1962. (Courtesy of the National Aeronautics and Space Administration.)

sometimes, it is thought, the solar wind flows directly into the ionosphere. The lack of a significant magnetic field on Venus is consistent with its exceedingly slow rotation. Supposing that Venus has a hot fluid core like the Earth's, we find no rotation to induce the inner motions that produce the Earth's magnetic field.

Hitherto in our survey of the solar system we have found planetary exteriors that are completely undesirable for home sites. Pluto is too cold; Mercury is both too hot and too cold; and both planets lack atmospheres, as do most satellites. The giant planets, as we shall see, are covered with noxious or poisonous gases and may possibly lack solid surfaces. Here in Venus we have an Earth-like planet, but it is too hot and has no oxygen.

11

Mars

Mars was named for the god of war because of the planet's sanguine color, obvious to the naked eye and more conspicuous with a telescope. The name, unfortunately, was much too appropriate during a number of years near the turn of this century. An astronomical battle was raging at that time and Mars was the battlefield. On one side was Percival Lowell, who carried on the banner first raised by Schiaparelli. On the other side stood a considerable fraction of the astronomical world. The *casus belli* was the observation of "canals" on Mars by both Schiaparelli and Lowell, and Lowell's interpretation of these narrow markings as artificial waterways. Schiaparelli used the Italian word *canali*, which means primarily *channels* or *grooves*, and did not believe that the canals were artificial. Lowell based his interpretation on his own extensive observations of Mars. Some of his composite drawings are shown in Fig. 135.

In the scientific world disagreement among authorities contributes to real and substantial progress. Usually the contenders

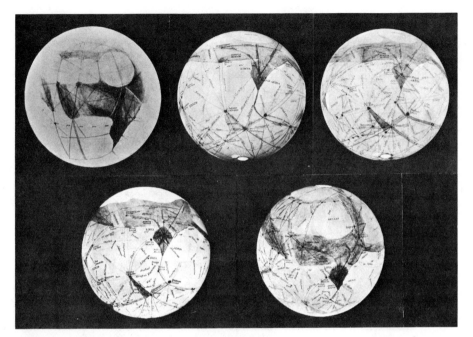

Fig. 135. Lowell's maps of Mars for the years 1894, 1901, 1903, 1905, and 1907, drawn on globes and photographed (south is up). (Courtesy of the Lowell Observatory.)

are each partially right and each partially wrong, but the heat of discussion furthers observation, which is the foundation of science. The Martian battle is over and the smoke has cleared. We can hardly say that Schiaparelli won a decided victory, but Lowell clearly lost. Nevertheless, astronomy gained because Lowell's enthusiasm stirred imaginations and spurred observations.

Today we know Mars better than we knew the Moon before the space program, thanks largely to NASA. In 1965 Mariner 4 showed us views of Mars, nearly comparable to Earth-based photographs of the Moon, while the Viking Landers and Orbiters in 1977 brought us right to the surface.

When Mars is most favorably situated for observation, a magnification of some 70 times enlarges the disk to the apparent diameter of the Moon. Small telescopes can be used satisfactorily at such a magnifying power, while larger ones are efficient at much greater powers. Since considerable detail on the Moon is

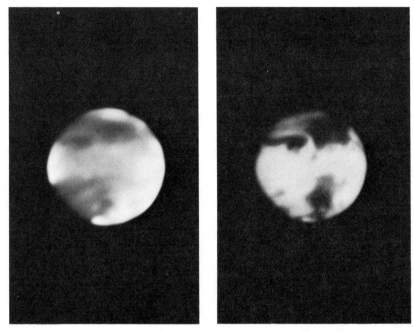

Fig. 136. Photographs of Mars with the 200-inch reflector: (*left*) in blue light; (*right*) in red light (south is up). (Photographs by the Mount Wilson and Palomar Observatories.)

visible to the naked eye, the reader may wonder why Mars should be difficult to observe. The difficulty is again the seeing discussed in Chapter 10. Under the best observing conditions on the ground the eye can occasionally resolve contrasting points 30 to 50 kilometers apart at the minimum distance of Mars. Very large telescopes do only a little better than those of aperture 50 to 80 centimeters (Fig. 136). Photographs generally resolve no better than about 300 kilometers. Even great mountains or valleys cannot be detected on Mars because no shadows can be seen when Mars is best placed for observation, directly opposite the Sun.

Even as first observed with a small telescope under very ordinary conditions of seeing, Mars immediately gains an individuality. Lowell wrote: "Almost as soon as magnification gives Mars a disk, that disk shows markings, white spots crowning a globe spread with blue-green patches on an orange ground." This verbal picture of Mars is somewhat more striking than the sensory registration of a novice who first observes Mars with a small telescope under average conditions. But by persistent observation, night after night, his eye will become more and more expert until he is able to distinguish surface details that were at first completely invisible. This remarkable improvement of visual acuity with experience has sometimes been underestimated, even by skilled observers who have not concentrated on planetary observations.

Mars can be well observed at intervals of about 2 years and 50 days when it comes into opposition with the Sun. Its synodic period with respect to the Earth is 780 days, about 50 days longer than 2 years (see Appendices 2 and 3 for the geometry and numerical data). Its distance from the Earth at opposition varies by a factor of nearly two (from 55,700,000 to 101,200,000 kilometers) because of the high eccentricity of the Martian orbit. The most favorable oppositions for observation are, of course, those at which Mars is the nearest, that is, when opposition occurs at perihelion. Since oppositions occur successively later by 50 days in alternate years, a favorable opposition will be repeated in seven or eight periods, at intervals of 15 to 17 years. The perihelion of Mars' orbit is so oriented that Mars is always best located for observation near August (1956, 1971, 1988). The positions of Mars at various oppositions are shown in Fig. 137.

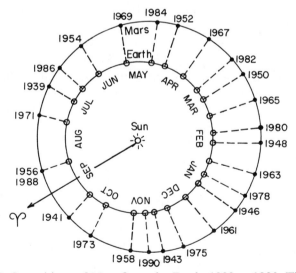

Fig. 137. Oppositions of Mars from the Earth, 1939 to 1990. The relative distances are shown by the lines joining the orbits. The seasonal dates on the Earth are indicated. Mars is north of the equator for oppositions from September to March.

The equator of Mars, like the Earth's, is tipped some 24° to the plane of its orbit, the direction of the axis remaining fixed in space. Consequently, at the times of Mars' closest approach we always see the planet in the same relative position. The south polar cap, by chance, is the one best observed, the north polar cap being turned toward us at the less favorable oppositions.

Mars rotates on its axis in 24 hours 37.4 minutes, making its day, from noon to noon, nearly 40 minutes longer than our day. The rotation may be noticed after less than an hour's observing. On the succeeding night the planet presents the same side because it has completed almost one turn in the meantime. In a little over a month the Earth gains a whole rotation, sufficient to complete a cycle of observations entirely around the planet. Similarly, in 24 hours, observers distributed around the Earth can observe the total circumference of Mars under optimum conditions of phase.

Photographically or visually the polar caps are usually the most conspicuous markings on the planet. The seasonal changes, first noted by Sir William Herschel, are regular, and even predictable with considerable accuracy. As the autumn season gives way to winter on one hemisphere of Mars, the corresponding polar cap grows irregularly until it may extend nearly halfway to the equator, to latitude 57° in the northern hemisphere and 45° in the southern hemisphere, the latter being the colder in the winter but the warmer in the summer. With the coming of spring (in the Martian March), the cap begins to recede; by the end of the Martian July it has disappeared at the south pole; the northern polar cap never quite disappears. In Fig. 138 the photographs present the same side of the planet to show the recession of the south polar cap. The seasonal dates for Mars are taken to correspond to seasons on the Earth; it must be remembered that the Martian year is 687 days, nearly equal to 2 Earth years. From a careful inspection of the series of photographs we can see the general darkening around the white area as the cap wanes in March and May, the intensification of the dark areas progressively away from the pole in June and July, and their fading by August. This sequence of events repeats with local and systematic variations every Martian year.

Even small details in the surface markings will reappear at the same season in different years. A detached area of the south

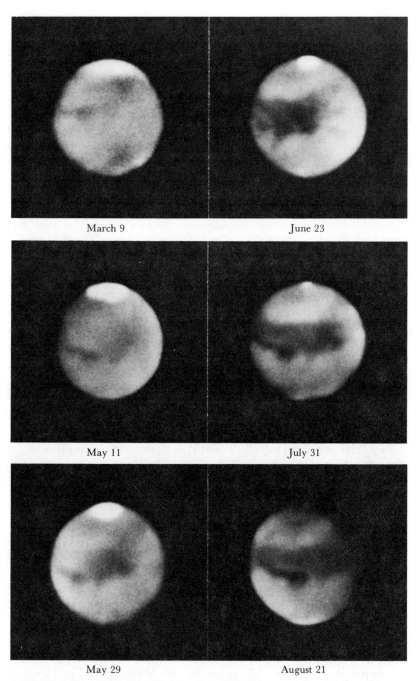

March 9	June 23
May 11	July 31
May 29	August 21

Fig. 138. Seasons on Mars. The dates given are Martian seasonal dates, taken to correspond to those on the Earth (south is up). (Photographs by E. C. Slipher, Lowell Observatory.)

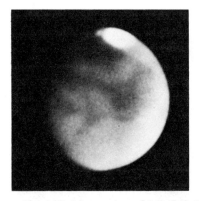 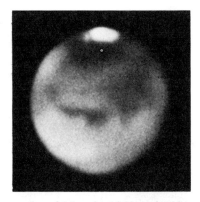

Fig. 139. Mountains of Mitchel. Photographs of Mars in 1909 and 1924. The detached area of the Martian polar cap appears at about the Martian date June 3 (south is up). (Photographs by E. C. Slipher, Lowell Observatory.)

polar cap is visible in the two photographs of Fig. 139. The first photograph was taken in 1909 and the second in 1924, but both on the Martian date June 3. The persistent area bears the name Mountains of Mitchel. The Viking Orbiter 2 pictures settle a long-standing question; the Mountains of Mitchel are a high "peninsula" (Fig. 140), not a depression. The great impact basin below them in the picture is almost frost free.

The repetitive character of the changes in the polar caps suggests immediately that these white areas are snow, which melts as the temperature rises. An alternative material is carbon dioxide or "dry ice." As we shall see, the answer is a mixed one.

As a polar cap begins to form in the Martian autumn, variable bluish-white clouds can be observed. The first two photographs in Fig. 141 were taken on successive nights in 1939 by E. C. Slipher (1883–1964). A white cloud near the north pole (*left fig., bottom*) has disappeared by the next night (*middle*). Another cloud is present six nights later (*right*). These clouds persist during the growth of the polar caps and special observing techniques are required to distinguish them from the frost or snow on the ground. They have been repeatedly observed by the spacecraft.

Infrared light can penetrate haze and dust in the Earth's atmosphere where blue or violet light will be stopped. Figure 142 presents photographs taken by W. H. Wright (1871–1959) at the Lick Observatory. At the top are Mars and the valley of San José as photographed by violet light, while at the bottom infra-

Fig. 140. Mosaic of Mars' south polar cap by Viking Orbiter 2, 1977. The Mountains of Mitchel are seen in the long "peninsula," upper right. (Courtesy of the National Aeronautics and Space Administration.)

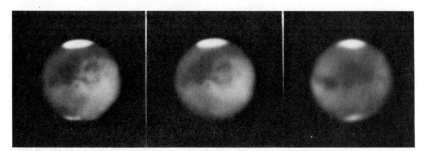

Fig. 141. Clouds at the north pole of Mars, 1939: (*left*) cloud near the north pole, bottom; (*middle*) the cloud has vanished on the next night; (*right*) another cloud 6 days later (south is up). (Photographs by E. C. Slipher, Lowell Observatory.)

Fig. 142. Mars and valley of San José, as photographed from Lick Observatory. Violet light was used in the upper photographs and infrared light in the lower ones (south is up). (Photographs by W. H. Wright, Lick Observatory.)

red light has been used. On Mars dust is apt to be the major culprit in obscuring the surface from external view, whether from the Earth or from near-Mars space. The Mariner 9 mission, in November 1971, barely succeeded, being threatened by a violent planet-wide dust storm that developed while the spacecraft was enroute. The storm fortunately abated in the nick of time. Such dust storms are more frequent when Mars is near perihelion.

The Earth-based observations of Mars have been highly successful with regard to the polar caps and their variations, dust storms, clouds, temperature, and color. But the amazing topography of Mars was not even suggested before the space age, radar having developed concurrently. On Fig. 143 the shaded areas and names depict the best representative map of Mars made before the Mariners. The features are identified by their Latin names. *Mare* means a sea, *sinus* a bay or gulf, *lacus* a lake, *lucas* a grove or wood, *fretum* a strait or channel, and *palus* a swamp or marsh. Note the hope expressed by these names. The brighter regions have unqualified names such as Elysium or Hellas. Various observers agree rather well in assigning differ- ent colors to the various regions, but most modern observers of Mars agree that any greenish tints observed arise from a physio- optical illusion. The canals are clearly the result of the eye-brain attempt to model lines and patterns from a mottled, irregular, diffuse object.

We can now compare the Earth-based picture of Mars with a geologic-type map based on all the Mariner pictures (Fig. 144). The strongest common feature is the great Hellas basin, fol- lowed perhaps by the Isidis Regio half circle next to Syrtis Major and then by the Coprates dark streak now recognized as part of the great Valles Marineris. Other darker areas in the Earth view are confirmed but none give even an inkling of the true surface character of the planet. In William K. Hartmann's inspired representation, Fig. 143, we see the distribution of cra- ters with diameters greater than 64 kilometers. Note how they cover roughly one hemisphere of Mars, the southern hemi- sphere if a north "geologic" pole is taken at longitude 155° and 55° north.

As we have seen, the number of craters per unit area is a mea- sure of the age of the peppered surface. On the Moon the old highlands carry the scars of ancient cratering while the later maria show few larger craters. The southern "geologic" hemi- sphere on Mars is thus the oldest hemisphere. But what has happened on the "new" hemisphere? A great deal! The plains of the new hemisphere are entirely different from the great basins of the near side of the Moon. They include the highest mountain peaks known among the planets, huge "shield" vol- canos nearly circular like Mauna Loa. In Fig. 143 their positions are marked with radiating spokes. The region of these major volcanos is called Tharsis, and represents a bulge on Mars about 1.2 kilometers above the average surface level measured from the center of Mars. With the deepest depressions approximately 6 kilometers lower than the average, the total effective elevation difference on Mars is about 27 kilometers, compared to about 19 on Earth and less on the Moon and Venus(?). The Tharsis bulge, however, is much smaller than the equatorial bulge aris-

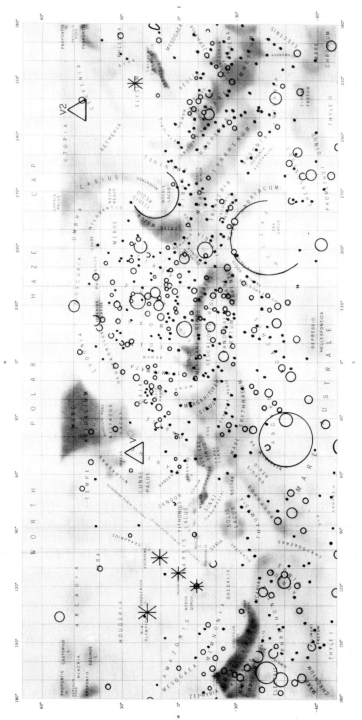

Fig. 143. Chart of Mars to latitude ±60°. The shadings and names represent the map of Mars as seen from the Earth. The circles and dots are the craters of diameter greater than 64 kilometers. The Viking landing sites are marked with triangles and the great volcanos with radiating spokes. (Courtesy of William K. Hartmann.)

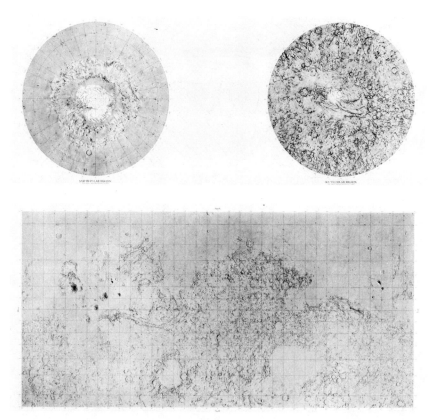

Fig. 144. The Mariner 9 "geological" map of Mars and polar regions. (Courtesy of the Jet Propulsion Laboratory, U.S. Geological Survey and National Aeronautics and Space Administration.)

ing from the rotation of Mars (about 17 kilometers). Figure 145 shows the general topography of Mars, mapped by radar.

The majestic volcanos on the Tarsus uplift dwarf their terrestrial counterparts. The greatest, Nix Olympia or Olympus Mons, towers some 21 kilometers above the surrounding plains, more than twice the elevation of Mount Everest above sea level (8.8 kilometers). The multiple caldera on the summit of Olympus Mons or Nix Olympia (Fig. 146) look superficially much like the caldera of Hawaiian volcanos such as Kilauea, but the former extends for a maximum diameter of 80 kilometers, most of the extent of the largest Hawaiian island.

Arsia Mons has an even larger crater, some 125 kilometers across. Associated with the Tarsus volcanos and uplift are huge

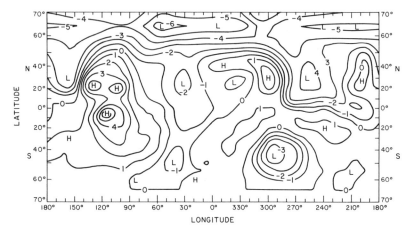

Fig. 145. Topography on Mars mapped by radar. Because of the coarse grid, details such as the great volcanos do not show. The contour interval is 1 kilometer. (Courtesy of James B. Pollack.)

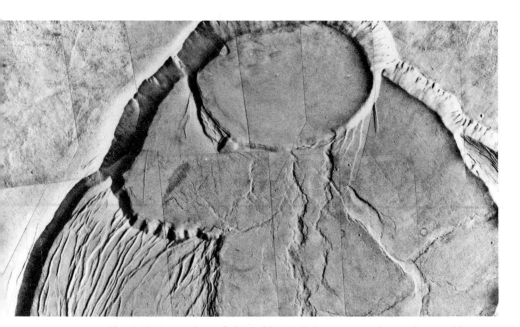

Fig. 146. A portion of the caldera of the great extinct volcano, Olympus Mons. The large crater (top center) is about 25 kilometers across with wall heights exceeding 2.5 kilometers. Note the rarity of impact craters, evidence for the relative youth of the volcano. (Courtesy of the National Aeronautics and Space Administration.)

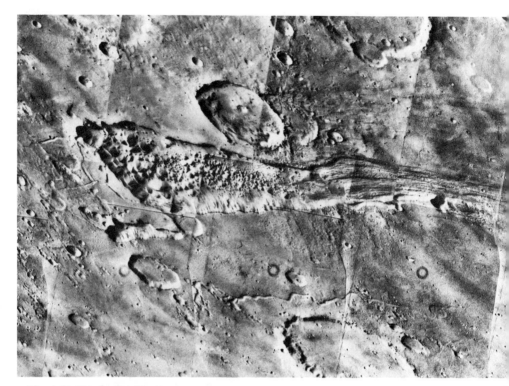

Fig. 147. Head of Valles Marineris or Coprates on Mars by Viking 1, July 3, 1976. The area covers 145 × 65 kilometers. (Courtesy of the National Aeronautics and Space Administration.)

systems of multiple fractures and ridges, some extending for as much as a 1000 kilometers and generally radiating from the central region of the great volcanos. These fractures and ridges attest to strains imparted to the surface by the Tarsus uplift.

In contrast to the uplift, volcanos, and lava flows, convection of the once molten interior of Mars also produced a magnificent rift valley, probably generically similar to the great oceanic rifts on Earth that encroach on the land in Ethiopia. One is visible as the Red Sea. The Valles Marineris or the Coprates Chasm lies just south of the equator and, with a great sweep, nearly parallels it for some 4000 kilometers, equal to the radius of Mars. The formation begins with a fractured distorted grid-like surface structure at its west end, southeast of the great volcanos. The chasm as such appears to begin with "chaotic terrain" (Fig. 147) and develops eastward into a colossal canyon sometimes

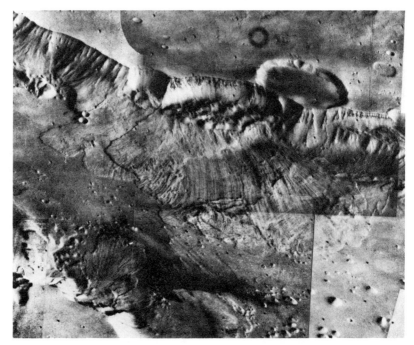

Fig. 148. A 70-kilometer section of Valles Marineris, the broad Gangis Chasma. (Courtesy of the National Aeronautics and Space Administration.)

nearly 200 kilometers wide, with many branching canyons as tributaries (Fig. 148). In some regions the depth of the canyon from the wall amounts to 10 kilometers, measured by the increased absorption bands in the carbon dioxide spectrum. It finally disappears in the Margaritifer Sinus, a ground-based dark area as seen from Earth. Before discussing the physical processes and time scales involved in the Martian topography, we shall return to questions of atmosphere, clouds, winds, temperature, dust storms, and so on, all of which are modified by the extraordinary gross structure of the planet.

The surface temperature of Mars was rather well defined at minimum by infrared Earth-based measures but overvalued at high noon. Viking Orbiter 1 provided the global temperature structure of the surface as seen in Fig. 149. The temperature rises to a maximum of −33°C near the subsolar point (upper right in Fig. 149). The morning terminator, around −120°C, shows clearly in the figure, where the contour intervals are re-

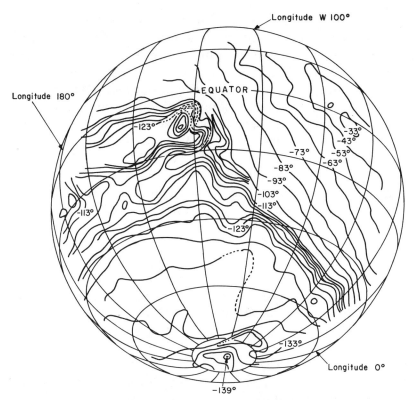

Fig. 149. Temperatures on Mars as measured by Viking Orbiter 1. Temperature contours are 10°C on the day side (*upper right*) and 2°C on the night side (*lower left*). (Courtesy of the National Aeronautics and Space Administration.)

duced from 10°C on the day side to 2°C on the night side. The great volcano Arsia Mons, elevation some 17 kilometers, grossly distorts the temperature contours just south of the equator near longitude 120° West. The east-facing slopes are in the sunshine while the west-facing slopes are still extremely cold, as is the peak, −130°C. To the east clouds or frost were seen after dawn. Generally the high Tharsis bulge area is colder while the great canyon areas and Hellas basin are warmer than the average for Mars. The lowest temperature is −139°C, near the south pole, where carbon dioxide can condense out. The maximum temperature measured at the Viking Lander 1 site is −28°C, although higher subsolar temperatures occur in certain of the equatorial regions. The temperature drops to about −130°C at altitude 40 kilometers and, with variations, remains at about that value at greater heights. The so-called *oases* in the Phoenicis Lacus (Solis Lacus) and Noachis sectors show daily temperatures of −53 to +22°C in the summer and −103 to −43°C in the winter. R. L. Huguenin and colleagues suggest that these regions are sources of water vapor for deposition on the north pole of Mars. Nevertheless we can conclude that Mars is frigid.

The amount of atmosphere on Mars is also most disappointing. The average pressure amounts to only 0.6 millibars or 0.6 percent of the terrestrial value. Because the surface gravity of Mars is only 0.38 that of the Earth, the mass of atmosphere per unit area scarcely exceeds 0.2 percent, and drops to about 0.07 percent of the Earth's value over the highest peaks on Mars.

The atmosphere of Mars, like that of Venus, is mostly carbon dioxide (CO_2, 0.95 by volume), nitrogen (N_2, 0.027), argon (^{40}Ar, 0.016), and oxygen (O_2, approximately 0.02; note the CO_2 band in Fig. 165). Although the carbon dioxide/nitrogen ratio is much like that on Venus, the argon abundance is *relatively* very greater but *absolutely* perhaps only 0.5 to 0.05. The water-vapor content is of great interest, particularly with regard to the nature of the clouds on Mars and to the possibility of life there. As might be expected, the water-vapor content is highly variable. The very maximum measured by the infrared spectrometers on the Viking orbiters was 0.10 millimeter of precipitable water. This occurred over the dark circumpolar region at the end of summer where the residual material appears to be dirty water ice. The vapor is saturated but still corresponds to barely 2 percent of the nitrogen content of the atmosphere. At this time the water-vapor content dropped with latitude from 0.10 millimeter at the pole to 0.015 at latitude 50°N to 0.005 at the equator and to less than 0.001 millimeter of water at latitude −40°. An "oasis" or "wet" spot occurred northwest of the Hellas basin in the Noachis sector. As the season advanced to the fall equinox for the northern hemisphere the water-vapor content increased near the equator and to the south. The total quantity of water vapor in the Martian atmosphere appeared to remain constant at about 1.3 cubic kilometers of water over the three Martian months of observation.

The question of seasonal water-vapor migration from pole to pole on Mars remains open, the variations with date being possibly a matter of temperature and surface characteristics only. In contrast, carbon dioxide clearly migrates. The total atmospheric pressure at the two Viking lander sites reached a minimum just before the autumnal equinox on Mars. It then rose by more than 30 percent to a maximum at the winter solstice. The minimum occurred just when the great southern polar cap had reached its maximum extent. The maximum pressure occurred near the time of the polar cap minimum. Because the atmosphere is almost all carbon dioxide and the southern polar cap shows temperatures suitable for the condensation of carbon dioxide, it appears that an appreciable fraction of the total atmosphere can freeze out, a total of 5000 cubic kilometers, according to S. L. Hess and his colleagues. This corresponds to 23 centimeters of deposition on the southern polar cap, neglecting some carbon dioxide that moved to the northern polar cap, reducing the atmospheric content at the time of the northern winter solstice.

But why so much more carbon dioxide moving in and out of the south cap? Perhaps because it is farther from the Sun than the north cap in midwinter. But two other factors are relevant: concentration of water to the northern hemisphere and north polar regions and the global circulation of dust. The great dust storms occur near perihelion in the southern summer so that systematically the warmer southern summers drive the water to the northern hemisphere where it is trapped in the dust as soil or permafrost. The remarkable sand dune structures around the north polar cap (Fig. 155) suggest the dust tends to accumulate more in these regions, although one of the most striking pictures of sand dunes (Fig. 150) occurs in the Hellispont region at 48° S and 33° W. Because the dust storms tend to occur during the northern winter, the dust also is carried to the northern polar regions. The accumulation there of permafrost dust is not directly measurable but may amount to kilometers.

We may summarize the situation in simple terms. The atmosphere *is* carbon dioxide. The polar temperatures are extreme enough to condense it and to sublimate carbon dioxide with changing seasons. Water is scarce in the cold atmosphere and therefore not as mobile planetwise as the carbon dioxide. Since the summers at the north pole occur near aphelion, the north polar regions are statistically colder than the south polar re-

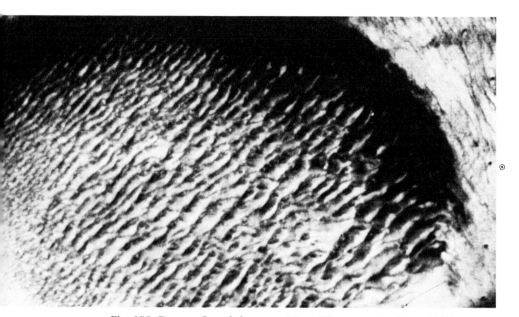

Fig. 150. Desert of sand dunes on Mars. The system is about 50 kilometers long so that the peaks of the dunes are separated typically 1 to 2 kilometers. By Viking 1 Orbiter. (Courtesy of the National Aeronautics and Space Administration.)

gions. Consequently, the water collects systematically at the north pole. The dust storms near perihelion carry dust to the north pole, where it settles with the condensation. At this time the south pole is sublimating so that the rising gas tends to discourage the accumulation of dust. Although other factors enter the process, the apparent result is to make the south polar cap of Mars primarily carbon dioxide and the north polar region a more permanent ice–dust accumulator.

The "oases" theory even suggests that water is carried underground from the north polar regions to be evaporated from the oases. The major support of the theory is the difficulty of initiating dust storms in the very low density atmosphere of Mars. Perhaps, then, the dust is catapulted into the atmosphere by evaporation of brine during the "heat" of the southern equatorial Martian summer, thereby accounting for the prevalence of dust storms at that season. No measures have been made of the temperature gradient with depth in the crust of Mars. There is enormous evidence that permafrost is ubiquitous. The tempera-

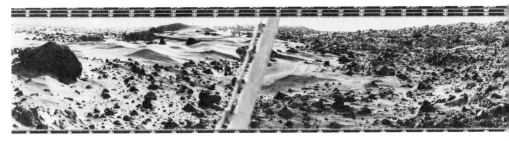

Fig. 151. Viking Lander 1 panorama. 100° angle of view. Wind-blown dust and dunes are apparent. The large boulder to the left measures about 1 × 3 meters at 8 meters distance from the lander. The Viking boom shows near the center. (Courtesy of the National Aeronautics and Space Administration.)

ture must surely increase with depth from radioactive heating as it does on Earth and the Moon. Thus a possible underground water table and underground water movements bear at least some consideration. Are we rational in shifting the "canals" on Mars from the surface to subsurface? Probably not, but the idea is intriguing.

The prevalence of dust on Mars shows in the magnificent close-up pictures sent back by the Viking Landers (Figs. 151 and 152). A reddish color is evident when color reconstruction is used. Thus our ground-based observations have not deceived us as to the color of the "God of War." Except for the color, such scenes are frequent in the Sahara Desert. Probably the sur-

Fig. 152. First close-up picture of Mars by Viking Lander 2, September 1976. The rocks are about 10 to 20 centimeters across. (Courtesy of the National Aeronautics and Space Administration.)

face rocks are more numerous near the Lander sites than the average for Mars because heat retention, or *thermal inertia,* is higher there than for 80 percent of the total surface. The visibly darker regions of Mars show a higher thermal inertia than the brighter regions. Hence the dark regions contain more rocks or bonded particles while the bright regions are covered more with loose fine dust, fewer rocks or bedrock being exposed. The Chasm Marineris and its "outlet" regions to the east and north stand out conspicuously on a global plot. They are areas of high thermal inertia and are therefore relatively rocky or consist of bonded particles. The dust-covered plains, of course, show a low thermal inertia in contrast to the volcanic plains and plains with ice in the ground. At the Viking lander sites the surface material varies from very compressible fine sand to clods and hard rocks. During the winter of 1979 water snow (frost?) condensed visibly on the surface around Viking Lander 2 and remained for months (Fig. 153).

Fig. 153. Snow on the region of Viking Lander 2. The covering is perhaps less than $\frac{1}{10}$ millimeter thick, possibly of water snow precipitated by carbon dioxide condensing on extremely small ice-dust particles. The carbon dioxide later sublimates away. (Courtesy of the National Aeronautics and Space Administration.)

Dust is second only to the polar cap variations in changing the appearance of Mars. Despite the low density of the atmosphere and necessarily high winds required to move dust on Mars, the dust is surprisingly mobile. The low gravity partially compensates for the lack of atmosphere, which, in turn, adds less resistance to a dust grain that has been raised. Hence the very fine grains have long paths. In falling back each grain knocks out other grains, the process of *saltation* whereby the global dust storms develop when the wind rises. Carl Sagan, Joseph Veverka, and collaborators find that winds from 50 to 90 meters per second are necessary to produce the observed wind streaking on Mars. Several large dust storms have been observed at the Viking Lander sites with associated temperature and pressure variations of great interest to meteorologists.

The white streaks in Fig. 154 are fine sand driven by the prevailing wind. This streaking provides an excellent means for charting the winds systematically with Martian seasons. The dark streaks result from scouring of the rocks by very high winds deviating some 50° from the prevailing winds. The darkening in Syrtis Major arises from huge areas of such dark streaks, presumably caused by wind scouring. In the North Polar region in the Martian spring, huge-scale terracing indicates the past episodic layering of permafrost sand (Fig. 155). We have no time scales for these episodes nor do we know the depth of these polar deposits. Generally speaking, however, these polar deposits and the large amount of dust accumulated in craters and crevices indicate that the great plains on Mars have largely been swept clear of dust. On Earth a far greater amount is available for wind transport should the desert areas increase by changes in climate.

The permanent magnets on the Viking landers were quite successful in attracting dust that was raised from the retro rockets, from sample acquisitions, and on the backhoe. This Martian loose soil contains 1 to 7 percent of highly magnetic material. Among a number of possibilities, the simplest seems to be that the red material contains maghemite, γFe_2O_3, a different crystalline form but the same composition as hematite. It may or may not have a covering of red ferric oxide. The color of Mars surely arises from a red iron oxide, the precise chemistry of which remains uncertain. The spectra of Mars suggest that the darker areas contain less oxidized basalt with more FeO

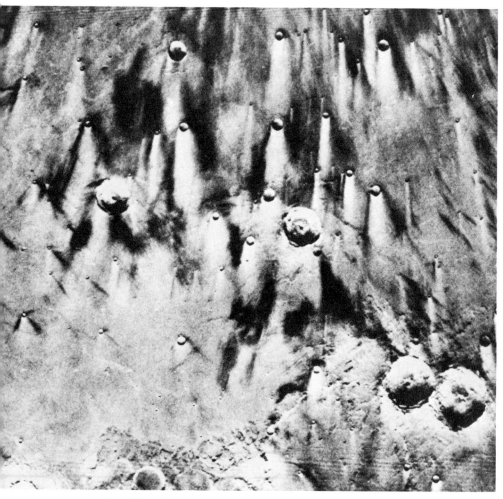

Fig. 154. Wind streaks behind craters on Mars. White streaks are fine sand deposits and dark streaks thought to be underlying rocky terrain scoured by high winds. (Courtesy of the National Aeronautics and Space Administration.)

than the brighter areas in which ferric iron and perhaps clay products from weathering are more prevalent. Sulfur may occur as a cement between particle surfaces.

Mars does not have a strong magnetic field although the magnetic measures are sparse and do show a bow wave. Probably the magnetosphere interacts directly with the ionosphere much as it does for Venus.

Fig. 155. Near the North polar cap of Mars in midsummer showing patches of residual water ice, terraces of permafrost (sand and ice?) and sand dunes (right center). By Viking 2. (Courtesy of the National Aeronautics and Space Administration.)

Many meandering channels, tributaries, and flow structures in the broad channels (see Figs. 147, 148, 156, 157) suggest that huge water flows have occurred on Mars. Even though no liquid water is evident or likely at present, there is evidence for abundant permafrost areas. Could Mars once have possessed oceans or lakes? Probably not, because there a huge atmosphere should have left a considerable legacy of the heavier noble gases, which are not observed. If Mars were once volatile rich, we must postulate some global situation to have removed these rare gases, perhaps just early heating of the upper layers. The light gases, particularly hydrogen from dissociation of water, could of course have escaped rather rapidly. We can reasonably assume that once there was somewhat more water and atmosphere, though probably not enough to have produced rain, rivers, or lakes. But the present surface of Mars shows no evidence of the global erosional features that should accompany a widespread abundance of liquid water. We must restrain our hopes (prejudices?) that Mars was once an Eden. With such restraint we can entertain the real possibility that great episodic releases of water from the soil or polar caps may have produced the apparent flow markings on Mars.

Before discussing possible mechanisms for such events, let us

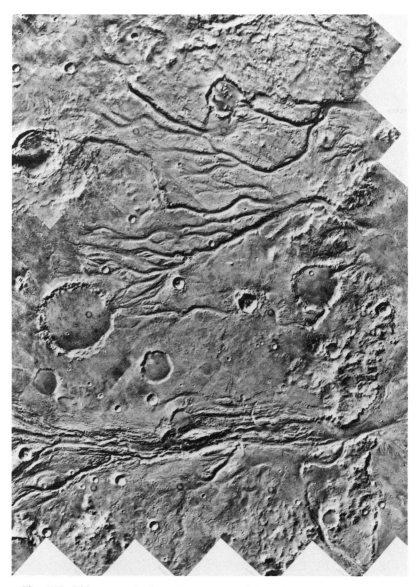

Fig. 156. Old stream beds on Mars, west of the Viking 1 landing site. The slope is to the right, some 3 kilometers in 140 kilometers across the picture. (Courtesy of the National Aeronautics and Space Administration.)

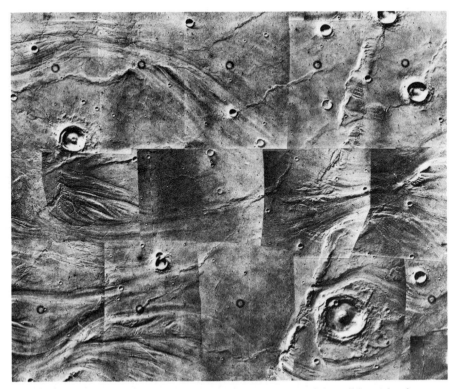

Fig. 157. Mosaic by Viking 1 of region west-northwest of the Viking 1 land-ing site showing lava flows and sinuous river channels. 200 × 250 kilometers in extent. (Courtesy of the National Aeronautics and Space Administration.)

look at an important property of Mars, the motion of its poles. Whereas the polar axis of the Earth precesses with a period of about 25,000 years (Chapter 7) because of the tilting force ex-erted by the Moon and Sun on its equatorial bulge (the spinning top effect), the polar axis of Mars precesses more slowly, in about 125,000 years. The Moon stabilizes the obliquity of the Earth's ecliptic so that it varies by only a degree or so from $23°.5$. The lack of a massive satellite allows the polar axis of Mars to wobble quite badly. William R. Ward calculates that the obliq-uity of Mars' equator to its orbit can reach extreme values of $35°$ and $15°$ centered about $25°.2$, close to its present value. The os-cillations occur with the period of precession, making the cycle about a million years (Fig. 158). We are now near a minimum in the oscillations. The effect of these obliquity changes on the

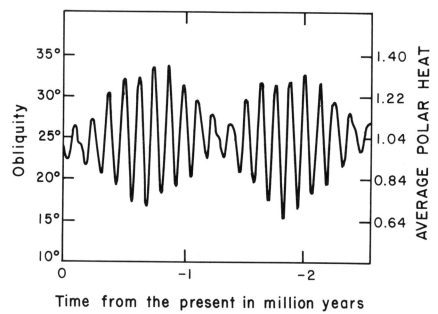

Fig. 158. Calculated oscillations in the obliquity of Mars' equator to its orbital plane (*left*) and corresponding average solar heat received by the poles (*right*), the present average being unity. (After William R. Ward.)

heating of the polar caps is profound, as pointed out by Robert B. Leighton and Bruce C. Murray. At a maximum obliquity, about half a million years ago, the polar caps should have received an average of about one-third more solar heat than they do today but at minimum obliquity only about two thirds as much, these extremes being separated by about 60,000 years. The ice in the poles may well have melted at maximum obliquity and most of the entire atmosphere frozen out at minimum. Estimates vary as to the total available budget of water and carbon dioxide for the atmosphere of Mars. At the minimum the total atmospheric pressure may well have dropped to below 1 millibar. The estimates at maxima vary from 30 to a dubious extreme of 100 millibars, the estimates still giving Mars much less atmospheric cover than the Earth.

Attempts to infer Martian temperature and weather during the periods of extreme melting of the polar caps are largely guesses. Will the carbon dioxide produce a greenhouse effect to warm the planet? Is there enough carbon dioxide and water to make a profound change? Will higher winds and atmospheric density make Mars a dust bowl? Will the clouds and dust increase the albedo to reduce solar heating and thus cool rather than heat the planet? Are the time scales of around 10,000 years enough to relocate much permafrost? These and other questions make accurate predictions impossible but highly provocative. The situation at minimum obliquity can be better predicted. The atmosphere will be extremely thin, too thin to produce dust storms, while both polar caps will be larger than their extremes today. High-energy solar flare particles will beat on the surface. Mars will be quiescent, hibernating.

Remember that these cycles have been repeated thousands of times in the past. If Mars has lost a considerable fraction of its atmosphere we can, with some justification, visualize that rather catastrophic events may have occurred when Mars was younger. We can visualize quiescent times when much more water was frozen out in the poles and in permafrost. The planet may possibly have shifted about its axis of rotation to change the latitudes of the regions we observe today. With such seemingly reasonable postulates in mind we can more readily accept explanations of the great erosional features on Mars. Consider the head of the great Valles Marineris in Fig. 147. Here we see truly chaotic terrain and even a 6-kilometer crater at its very tip. Suppose that at the end of a quiescent period of Mars this region was covered with a kilometer of permafrost encased in light material such as dust or volcanic ash. Underneath, a large amount of water may have been trapped in deeply buried *aquifers*. Planetary heat from the interior could have aided in the process. As the pressures on these aquifers increased, a small triggering event such as an impact may have set off a chain reaction that released a gigantic flow of water. The chaotic terrain in Fig. 148 then represents the result of undermining by the underground water release. At maximum several hundred millions of cubic meters of water per second are theoretically possible and are of the order of magnitude to produce the flow formations that show in several of these Martian channels (Figs. 148, 156, 157).

The speculative aspect of this catastrophic theory is greatly reduced by the fact that we have a terrestrial record of such an event. In 1923 J. H. Bretz presented evidence that the Chan-

Fig. 159. Aerial view of Dry Falls, which is 150 kilometers west of Spokane, Washington, and is the western edge of the Grand Coulee. Its features were cut by an unprecedented glacial flood some 20,000 years ago. (Courtesy of the U. S. Geological Survey.)

neled Scabland of eastern Washington (Fig. 159) was created by catastrophic flooding. This "outrageous" theory germinated for more than two decades before it was taken very seriously by the geological community. Apparently some 2000 cubic kilometers of glacial water were released at a rate of some 20 million cubic meters per second to produce the Scabland. The area of the chaotic terrain in Fig. 147 is over 900 square kilometers so that the release of a half-kilometer-deep layer gives us a somewhat comparable theoretical source for Valles Marineris.

The Viking close-up pictures of Mars show many bizarre features, many like those of the Moon and Mercury but others that are unique. No where else in the solar system have we seen a white rock like the one in Fig. 160. Some 14 × 18 kilometers in

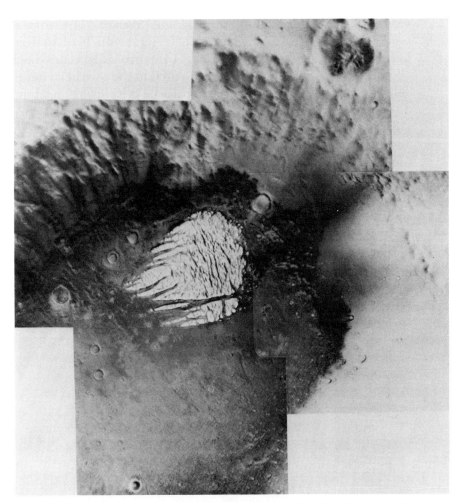

Fig. 160. The "White Rock" as imaged by Viking Orbiter in 1978. Latitude: 8° south and 135° west. 14 × 18 kilometers in size. (Courtesy of the National Aeronautics and Space Administration.)

size, this rock lies inside a 93-kilometer crater. Being near the equator of Mars, the rock cannot be made of ice or snow.

Our only direct dating technique for the features on Mars involves the crater counts. They are excellent for relative dating of large areas but uncertain for absolute dating. We have no precise knowledge of abundance of impacting bodies, Mars being nearer the edge of the asteroid belt than the Earth and Moon and very much nearer than Mercury. As a consequence, various investigators agree on the *relative* ages of Mars' formations but not on their *true* ages. In general, much of the old heavily cratered hemisphere (Fig. 143) probably dates back about 4 aeons. The crust that was formed in the northern hemisphere was probably destroyed by isostatic lowering at about this time. The Tharsus uplift probably occurred rather "rapidly" and became permanent in the interval 4.0 to 3.5 aeons ago, according to the best current opinion. The spectacular volcanos were most active and developing during this period. How long they remained active is in question; some investigators suggest that their activity is not yet complete, while the majority lean towards termination about 2 aeons ago. Isotopic studies of samples are an obvious need, suggesting a realistic space mission, whether by sample return or by *in situ* analysis.

Studies of soil samples from Mars' tiny satellites (Fig. 6) could both date and clarify their origin. A close-up view of Phobos by the Viking mission (Fig. 161) shows the irregularities expected

Fig. 161. Phobos, a 9.5 kilometer section. (Courtesy of the National Aeronautics and Space Administration.)

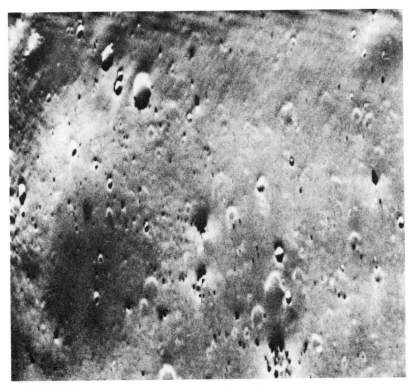

Fig. 162. Close-up of Deimos by Viking 2 Orbiter, only 1.2 × 1.5 kilometers across showing features as small as 3 meters. (Courtesy of the National Aeronautics and Space Administration.)

from saturated impact cratering. Not surprising was their rotation, like the Moon's, with the long axes pointing always toward their primary. Unexpected was the difference in surface structure between Deimos (Fig. 162) and Phobos, and the clear evidence of layers or rill-like depressions on Phobos, associated with the Stickney, the 8-kilometer crater with one-third the diameter of the satellite. The depressions or grooves, 100 to 200 meters across, may have resulted from a nearly catastrophic impact, from tidal stresses, from primary faults or layering, or from crater chains which are also observed. No definitive answer is yet available.

Deimos (Fig. 162) clearly shows a much deeper regolith than Phobos. A layer of dust seems to cover craters of diameter less than 50 meters across while house-sized rocks appear here and there. Whether the major differences in surface structure are inherent or arise from different crater-forming environments is not clear. Both satellites are uniformly gray in color and reflect like carbonaceous chondrites, the most primitive type of meteorites.

The suggestion that Deimos and Phobos are captured asteroids gains little support from celestial mechanicians, although rather unlikely scenarios cannot be disproved. More likely the satellites were formed in a ring around Mars, separately or possibly with others long lost. Perhaps they are fragments of a larger satellite, fissioned by a collision. This hypothesis gains support by theoretical evidence that Phobos, being the larger, should be drawn toward Mars by tidal friction more rapidly than Deimos, the smaller. Hence Phobos, now moving within the orbit of Deimos, must have crossed the latter's orbit. The more we learn about such bodies, the more penetrating are the unanswered questions.

A major goal, or perhaps *the* major goal, of the Viking missions to Mars was to search for extraterrestrial life forms. Three difficult biological experiments were performed: Pyrolytic Release, Gas Exchange, and Labeled Release. They are based on experience with terrestrial organic life.

The Pyrolytic Release experiment assumes that atmospheric carbon and possibly photosynthesis are involved. The soil sample is incubated for five days in an atmosphere consisting of radioactive carbon-14 in carbon monoxide and dioxide, replacing part of the Martian atmosphere. Artificial sunlight illuminates the sample to encourage photosynthesis, a filter cutting out the ultraviolet to prevent confusing nonbiological effects. Heating to 625°C pyrolyzes any organic material incubated, which is trapped chromatographically by a helium stream. The radioactivity of the residual atmosphere and the trapped organic material, if any, are compared to see whether the radioactive carbon has been assimilated.

The Gas Exchange experiment assumes that water is involved in possible Martian biology which also requires a chemical exchange with the atmosphere. The soil sample is supported in a porous cup raised above the floor of the incubation chamber in an atmosphere of carbon dioxide, krypton, and helium at the Martian atmospheric pressure. For the first seven days the cup is held above a water solution of biological nutrients in the chamber. Test samples of the atmosphere are checked in a chromatographic column for hydrogen, nitrogen, oxygen, methane, and carbon dioxide with krypton as a standard. If there is no evidence for biological activity the liquid level can be raised to the porous cup.

The Labeled Release experiment also assumes that water is necessary in Martian biology, that the microorganisms assimilate organic molecules and ions as nutrients, and that they release gases containing some of the carbon from the nutrients. The dissolved radioactively labeled nutrients are added to a soil sample in an enclosed dark chamber with a Martian atmosphere. The radioactivity of the atmosphere is monitored to determine whether the postulated microorganisms excrete any carbon compounds.

All three experiments gave active results which can be mostly explained on the basis that the materials came from the Martian desert. The particles appear to have an iron oxide coating that develops in a thin atmosphere without appreciable humidity where ultraviolet radiation has been intense. The strong reactions observed in all but the labeled release experiments follow from the chemistry expected *without* the assumption of biological materials on Mars. Earth-based experiments are continuing to clarify the chemistry.

Were it not for the Viking molecular analysis experiment, not aimed primarily for biological material, we might still hold considerable hope for life forms in the desert sands of Mars. In this experiment crushed soil samples are heated to 200°, 350°, or 500°C for 30 seconds. Volatiles and pyrolysis materials are measured in a gas chromatograph column and then caught in a palladium separator for further analysis with a mass spectrometer. In the first Martian sample traces of known contaminants, methylchoride and freon-E components, were identified, proving the efficacy of the system. The eleven-member scientific team headed by K. Biemann report that "none of the five gas chromatograms obtained in the experiments . . . showed any indications of the presence of organic compounds indigenous to the two samples. However at 350° and 500°C considerable quantities of water were evolved." Limits on the abundances of several hydrocarbons and oxygen-nitrogen- or sulfur-containing compounds were usually less than 0.01 parts per million.

We may conclude almost certainly that life as we know it shows no traces on the desert of Mars, even though our own deserts and the soils of antarctica give consistent positive results by these methods. If life has developed on Mars we must look elsewhere to find it. The prospect is not very encouraging except for the clear evidence that water has been widespread on Mars, at least periodically. Will we finally discover life forms in the "oases" or in the bottoms of the great canyons such as the Valles Marineris?

12

Great Jupiter and His Satellites

From the small rocky terrestrial planets, we now turn our attention to the antithesis of everything earthlike—to the colossus Jupiter, whose surface presents a turmoil of never-ceasing cloud transmutations. We find this planet some 630 million kilometers away in space; its mass amounts to more than 300 Earths, while its volume exceeds that of our planet by more than a thousandfold. Through the telescope we can see Jupiter as a golden disk with dark and light bands roughly parallel to each other. Reddish or brown shades of color catch our eyes while we note the irregular cloudlike patches that break the uniformity of the bands (Fig. 8 and Plates I and II). The disk seems slightly elongated in the direction of the bands, and careful measures confirm our judgment; this diameter is greater than its normal by one part in fifteen. The difference between the polar and equatorial diameters is 0.7 the Earth's diameter.

Within an hour's observation the planet appears to have turned appreciably; in only 9 hours 55 minutes it will have made a complete rotation. On a succeeding night we find that the bands and surface markings are much the same as before, but that the details are slightly changed. Within a few weeks the band structure is considerably transformed though the general character of the markings remains unaltered (Plate I). Since the axis of rotation is perpendicular to the plane of the slowly changing bands, the bands result from great "trade winds" or atmospheric currents parallel to the equator.

The rotation of Jupiter is indeed rapid. Near the equator there is a zone where the rotation appears faster than in higher latitudes. There the period of revolution is some 5 minutes shorter, only about 9 hours 50 minutes. Periods between 9 hours 51 minutes and 9 hours 53 minutes are rarely observed, and then not persistently. The equator turns with a velocity of about 40,000 kilometers per hour. These rotation periods obviously apply only to the clouds. The radio noise from Jupiter shows that the pole of its powerful magnetic field is tilted some 10° to the pole of rotation. The magnetic pole rotates very uniformly in 9 hours 55 minutes 29.37 seconds, suggesting at least a pseudo-rigid body rotation underneath the clouds.

The consequent centrifugal force of rotation is sufficient, even when acting against 2.6 times the surface gravity on Earth, to flatten the sphere by the appreciable amount already noted. The amount of flattening, however, is not as great as would be expected were the interior of Jupiter similar to the interior of the Earth. This shows that the density increases with depth more rapidly than for the Earth. When we consider that the *mean* density is only 1.34 times that of water, we must conclude that Jupiter is composed largely of extremely light gases, mostly hydrogen and helium, much like the Sun. Comparison of Jupiter's internal structure with that of the other planets is extended in Chapter 14.

The existence of a very deep atmosphere is evidenced by the directly observed markings. Most startling of all the features on the Jovian surface has been the Great Red Spot, first noted in 1830. It appeared as a brick-red area, elongated some 50,000 kilometers (nearly 4 times the Earth's diameter!) in a direction parallel to the equator. It is still striking as an oval spot of varying color and character, but has never been as conspicuous as during the first few years after discovery. Figure 163 presents

Fig. 163. Jupiter photographed with the 200-inch reflector in blue light (*top*) and red light (*bottom*). Note that in red light the Great Red Spot has essentially disappeared. Ganymede and its shadow are also shown. (Photograph by the Mount Wilson and Palomar Observatories.)

the aspect of Jupiter and the Red Spot as photographed in blue and in red light. In blue light the Spot is conspicuously dark against the planet's disk. In red light the Spot has almost disappeared. A white object would show equally well in both colors. The impartial "eye" of the photographic plate therefore verifies the color of the Great Red Spot. The Great Red Spot does not rotate uniformly with the planet but drifts about considerably, having wandered as much as three revolutions from its average position on the planet calculated with a constant period. Such freedom of motion shows unquestionably that the spot is a floating disturbance. Also it appears to be dying away. The almost incredible pictures of the Jupiter system by NASA's Pioneer 11 and Voyager missions with their accompanying physical measurements are clarifying the complex and majestic meteorology of Jupiter's atmosphere. But first let us look to the composition.

In Jupiter's atmosphere the molecules of ammonia (NH_3) and methane (CH_4) are responsible for the losses, dark bands, shown by the spectra of Fig. 164. Methane predominates in the spectra of other giant planets. Other dark lines to be seen in

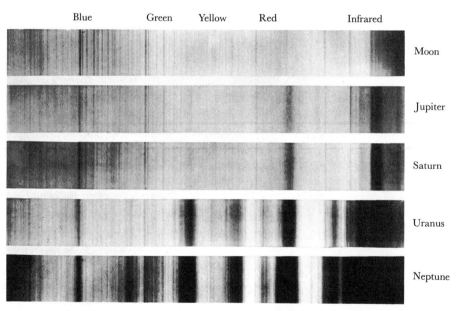

Fig. 164. Spectrograms of the Moon and the giant planets. The great dark absorption bands, most conspicuous in Neptune's spectrum, are due to methane. (Spectra by V. M. Slipher, Lowell Observatory.)

TABLE 4.   *Composition of Jupiter's atmosphere.*

Gas	Percent by volume
Hydrogen, H_2	90
Helium, He	10
Methane, CH_4	0.07
Ammonia, NH_3	~0.01
Ethane, C_2H_6	0.003
Water, H_2O	0.0006
Acetylene, C_2H_2	0.0005
Phosphine, PH_3	0.0005
Carbon monoxide, CO	Trace
Germane, GeH_4	Trace
Deuterium, D	As CH_3D

these spectra have been produced by the gases of the Sun's outer layer and by the Earth's atmosphere. Theodore Dunham of the Mount Wilson Observatory was able to identify these two gases by separately compressing them in a 20-meter pipe. He discovered that the wavelengths lost from a light beam reflected twice through the pipe agreed identically with the wavelengths absent in the spectrum of Jupiter. Some 10 meters of ammonia gas at standard atmospheric pressure are equivalent to the average amount in Jupiter's atmosphere, to the depth that sunlight penetrates before it is reflected back to us. For methane, the corresponding amount is about 160 meters. Continuing ground- and space-based observations now tell a great deal about the composition of Jupiter's atmosphere (Table 4). As expected, hydrogen and helium predominate. Even methane and ammonia are little more than trace components, even though they predominate in the infrared spectrum (see Fig. 165). Water has frozen to sink out of sight, leaving a barely perceptible amount as gas. The mean molecular weight of Jupiter's atmosphere is only 2.2 times the hydrogen atom, confirming ground-based estimates from the rate that stars are obscured as they are occulted by the limb of the planet. Above the clouds, atmospheric pressure falls by a factor of 2 in 10 to 12 kilometers. The light gases are compressed by the great gravity. The temperature there is dropping rapidly with height, from −113°C at 1.0 Earth atmospheric pressure, or 1 bar, to −160°C at 0.03 bar.

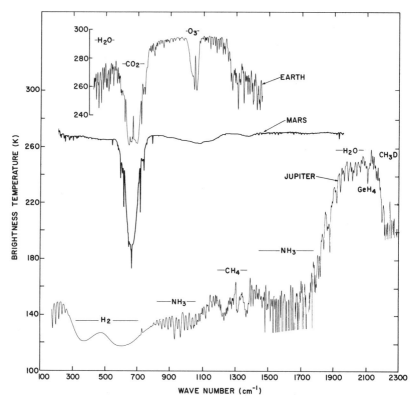

Fig. 165. The infrared spectrum of Earth, Mars, and Jupiter. The effective temperature is given in degrees Kelvin. Subtract 273° to obtain °C. The wave number is the inverse of wavelength, 1000 corresponding to 10 micrometers or 20 times the wavelength of blue-green light. (Courtesy of Robert A. West, Laboratory for Atmospheric and Space Physics, University of Colorado.)

The relative heights of features in Jupiter's atmosphere can be compared directly in Fig. 166. The image in the near-infrared dark band of methane is brightest over high regions where the methane layer is thin. Dark regions are deeper. An image in the nearby continuum of the spectrum gives a quick reference for individual features. Note that the clouds over the Great Red Spot and the equatorial band are quite high in Jupiter's atmosphere. Especially conspicuous are the poles, both showing the scatter from the haze layer, very high in these colder regions of Jupiter.

To understand the meteorology on Jupiter we must first recall that on Earth the clouds, winds, and circulation patterns are

Fig. 166. Jupiter imaged (*left*) in the middle of the dark infrared band of methane and (*right*) in the nearby continuum. (By Robert A. West using the University of Arizona's 61-inch telescope.)

fed by the energy of solar heat, introduced mostly at the surface and in the lower atmosphere. For Jupiter, heat measures show that the planet itself internally generates and radiates about twice the solar contribution. Furthermore, there is no solid surface or topography on Jupiter. The internal heat is carried by uprising convection of warmer volumes of gas, creating turbulence or eddies. On Earth we have the major heating in the equatorial regions and upwelling of the atmosphere there. Because of the direction of rotation these currents are moving easterly when they come down at the middle latitudes where the rotational speed of the surface is lower; hence the prevailing westerly winds. Nearer the equator we have the trade winds fed from higher latitudes and moving from east to west, that is, easterly trade winds. In a cyclonic disturbance the air is drawn toward a low-pressure area. It rotates counterclockwise in the northern hemisphere and clockwise in the southern because the air moving toward the equator deviates to the west, and that away from the equator toward the east. Hence they produce the typical cyclonic rotations as they converge. This oversimplified account of terrestrial circulation patterns gives some inkling, at least, of the complex Jupiter system.

Looking to Fig. 167 we see the various zones about Jupiter's equator. For about ±9° in the equatorial zone the winds are strongly west to east (westerly), approximately 100 meters per

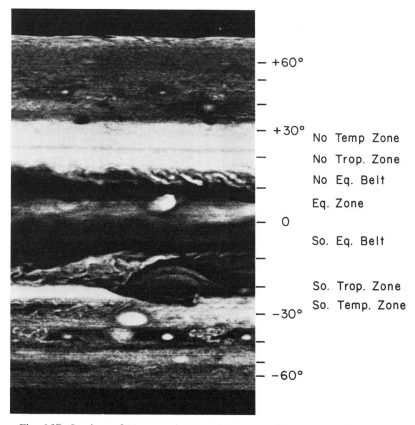

Fig. 167. Section of Voyager image of Jupiter with latitude of zones and belts indicated: Eq. = Equator; Trop. = Tropical; Temp. = Temperate. (Courtesy of the National Aeronautics and Space Administration.)

second, reversing to mild "trade winds," east to west (easterly) near latitudes ± 20° at about 50 meters per second. The Great Red Spot is carried west along with the South Tropical Zone. Farther from the equator we again find the westerlies, narrow with high speed, in the Temperate Zones, the great white bands. The North Temperate Zone is wider, more uniform, and moves more rapidly than the South Temperate Zone in the Voyager pictures. This condition varies and even reverses over the years. Note the striking long-term changes shown in Fig. 168.

The eddies and jets between the main flows are conspicuous. The Great Red Spot has no roots in the planet. We observe that

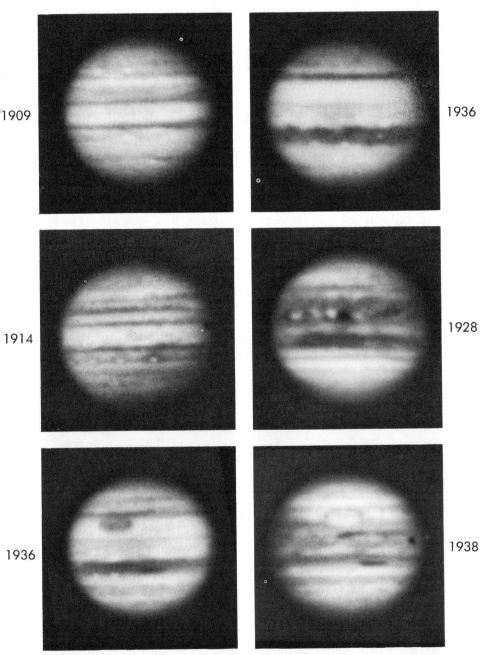

1909

1936

1914

1928

1936

1938

Fig. 168. Long-time changes on Jupiter. The photographs were taken in the years indicated. Shadows of satellites appear on the 1914 and 1938 photographs. (Photographs by E. C. Slipher, Lowell Observatory.)

the upper regions are rising and spreading away from the center, which accounts for the spot's low temperature and anticyclonic rotation, counterclockwise in the southern hemisphere. The rotation period in the spot is about 7 days. The great ovals are similar disturbances, now shrinking after their appearance in 1939. The Great Red Spot appears also to be shrinking, its lifetime being uncertain, possibly having begun before the spot was definitely identified around 1830. Meteorologists now hold that the Red Spot and White Ovals are *not* the result of catastrophic events such as volcanos or meteoritic impacts on Jupiter. They must evolve, grow, and be sustained by convection cells that carry the internal energy of Jupiter to the surface where it can be radiated away. Lapse-rate motion pictures of Jupiter made by the Voyager cameras show the remarkable eddies around the Great Red Spot as it appears to plow its way through the adjacent cloud structures. Some survive the encounter and others are drawn into wisps and disappear. The Voyager missions have given atmospheric scientists a gold mine of data and a ringing challenge for detailed theory.

The particles of the clouds are primarily ammonia, the temperature being in the proper range, -100 to $-160°C$. At one atmospheric pressure ammonia boils at $-33°C$ and melts at $-78°C$, so that it easily freezes out in Jupiter's atmosphere, but with a small vapor pressure, enough to produce the spectroscopic bands. Methane, on the other hand, boils at $-161°C$ and melts at $-184°C$, so that the liquid or crystalline ammonia is almost entirely excluded.

The cloud tops on Jupiter range in height over about 12 kilometers, much like the range for Earth clouds. We rarely observe to depths corresponding to more than two Earth atmospheres, in the temperate zones. Thus, the pressure range for Jupiter's clouds as seen in Fig. 166 is also comparable to that for Earth clouds.

The photographs portray distinct variations in shades and tints over the several belts and clouded areas. (Plates I and II). Stable atmospheric compounds cannot produce the colors seen on Jupiter. The colors must form, circulate, and disappear; otherwise the planet would assume a constant hue. Possibly colored metallic contaminations might be thrown up from below, say by volcanos, finally to settle back or be altered chemically by the atmosphere. Ruppert Wildt has suggested sodium as a possible contaminant, and Harold C. Urey has suggested that organic molecules color the clouds on the giant planets. Indeed, Carl Sagan and Stanley L. Miller have produced brightly colored organic molecules by spark discharges through a simulated Jupiter atmosphere in the laboratory. They suggest that lightning in such cloudy, turbulent, giant atmospheres produces chemically short-lived colored compounds. The Voyager missions have definitely identified powerful lightning flashes on Jupiter, comparable to the strongest on Earth. No relation between lightning and color is yet demonstrated, however.

The detection of phosphine, PH_3, on Jupiter suggests that the color of the Great Red Spot might arise from red phosphorus crystals formed when phosphine is decomposed by solar radiation. In any case the presence of phosphine, germane, and carbon monoxide in Jupiter's high atmosphere indicates strong vertical mixing by convection and eddies arising from great depths.

Radio astronomy has produced some remarkable new knowledge about Jupiter and raised some interesting questions. In 1955 B. F. Burke and K. L. Franklin of the Carnegie Institution of Washington were studying radio stars with an antenna system tuned to a wavelength of 13.5 meters when they discovered that Jupiter was also a radio star producing noise at semiregular intervals. C. A. Shain of Australia then searched records of Jupiter noise back to 1951 and discovered that the noise could largely be associated with a certain area of Jupiter's surface (see Fig. 169). Even though the noise is not always radiated at each

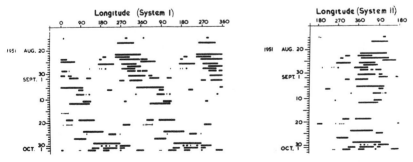

Fig. 169. Times of radio noise from Jupiter in 1951, as displayed by C. A. Shain. In the left-hand figure the times form a slanted structure because the observed surface period of rotation was adopted. The corrected period of rotation on the right brings the noise into synchronism with the solid(?) inner core.

rotation of Jupiter, certain longitudes do produce the noise systematically and indicate a solid-body rotation for the sources, with a period of 9 hours 55 minutes 29.37 seconds. The great radio-noise bursts from Jupiter correspond in energy to a billion simultaneous lightning flashes on the Earth and seem to be of very short duration, a small fraction of a second. They are hardly ever observed from the Earth at wavelengths beyond 20 meters because of Earth's ionosphere. Not surprisingly the Voyagers, in open space and nearer to Jupiter, detected very long wavelength noise and storms in the range 3 to 30 kilometers or 100 to 10 kilohertz. Their source appears to be associated with the inner part of the magnetosphere extending down possibly to the upper ionosphere of Jupiter and possibly with the inner satellite Io.

At microwave radio frequencies the temperature of Jupiter is close to the expected value from infrared measurements, $-143°C$, but as the wavelength increases the calculated effective temperature becomes much greater and indicates that Jupiter, like the Earth, has a magnetic field and radiation belt in which very high-frequency radio noise is continuously generated. The magnetic field around Jupiter is of the order of ten times that of the Earth and the magnetic axis, like that of the Earth, is tilted at a considerable angle to the axis of rotation, about 10 degrees. This, indeed, partially accounts for some of the semiperiodic character of the radio bursts at long wavelengths. The radiation in the decimeter wavelength region, above and below 3000 megahertz (3000 MHz), is clearly produced by relativistic electrons orbiting around Jupiter in its magnetosphere. They orbit and radiate best near the plane of the magnetic equator, hence defining the tilt and direction of Jupiter's magnetic pole from its rotation pole. The radio noise reaches maximum when the Earth lies in the plane of Jupiter's magnetic equator.

The Voyager missions delineate Jupiter's magnetosphere, much like the Earth's (Fig. 61) but enormously larger. The bow wave extends toward the Sun into the solar wind nearly 100 Jupiter radii (Rj) or 0.05 astronomical units, and the magnetosphere itself about 70 Rj or 5 million kilometers, embracing all of the Galilean satellites. Away from the Sun, the tail of the magnetosphere has been detected to approximately 170 Rj or 0.08 astronomical units.

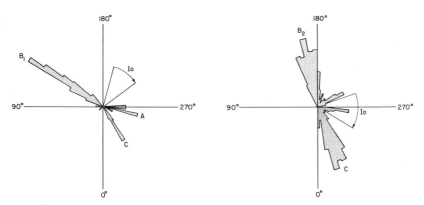

Fig. 170. Longitudes of Io (*left,* 195° to 235°; *right,* 250° to 300°) at which regions B_1, B_2, and C radiate 10-meter radio noise when they are on the central meridian as seen from Earth. The figure lies in the plane of the ecliptic with longitude in System III. (After C. N. Olsson and A. G. Smith.)

A mysterious and complicated structure of the magnetosphere became manifest in 1964. E. R. Bigg discovered that radiation in the 10-meter range occurred only at specific longitudes of the inner Galilean satellite Io. Figure 170 shows the amazing relation between the longitudes around Jupiter at which specific regions A, B_1, B_2, and C on Jupiter radiate when Io lies in two ranges of longitude. The radio-noise bursts seldom occur except when Io crosses the plane of Jupiter's magnetic field on one side of the planet as seen from the Earth. The other satellites seem to be scarcely, if at all, involved in the phenomenon, although the Voyagers find a nick in the magnetic field at Ganymede's distance from Jupiter.

The Io story began to unfold in 1973 when R. A. Brown discovered sodium emission lines in its spectrum. From Pioneer 10 observations D. L. Judge and R. W. Carlson found that Io was embedded in a toroidal cloud of hydrogen around Jupiter. Ground-based observations confirmed that the sodium cloud was indeed extended and that its brightness varied with position around the orbit because the sodium atoms radiate in proportion to the sunlight available for excitation. In their high-velocity motion around Jupiter the sodium atoms shift their absorption wavelengths with respect to the Sun's deep sodium lines by the Doppler effect (see p. 257). Voyager 2 found that the Jupiter plasma is increased by about 1000 electrons per cubic centimeter inside the Io torus. The radio noise connected with Io then arises from electrons orbiting in the disturbed magnetic fields induced in Jupiter's magnetosphere by Io and its accompanying plasma.

Jupiter's Ring and Satellites

Jupiter's satellite system has grown from four in 1610, when first seen by Galileo, to fourteen in 1979, plus an inner ring, newly discovered by Voyager 1 (Fig. 171). Seen with the Voy-

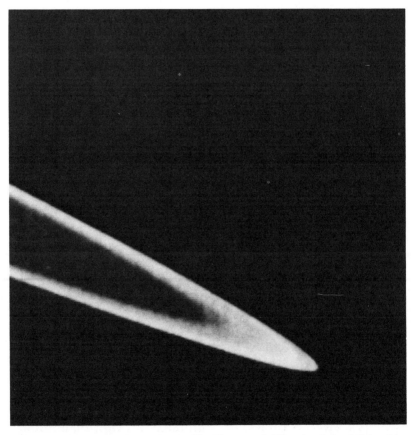

Fig. 171. Ring of Jupiter as imaged by Voyager 2. The ring is backlighted by the Sun and is divided near the outer edge (not obvious in the picture). A very tenuous ring extends almost to Jupiter's cloud tops. (Courtesy of the National Aeronautics and Space Administration.)

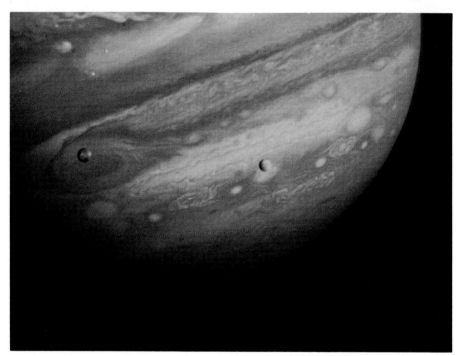

Plate I. (Top) Changes in the Great Red Spot hemisphere of Jupiter in four months, as viewed by (left) Voyager 1 and (right) Voyager 2. (Bottom) Voyager 1 view with the satellites Io (left) and Europa (right) seen against the disk. (Courtesy of the National Aeronautics and Space Administration.)

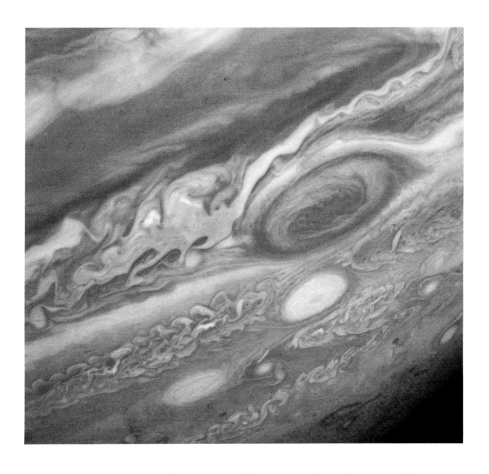

Plate II. Voyager 1 view of the Great Red Spot on Jupiter. (Courtesy of the National Aeronautics and Space Administration.)

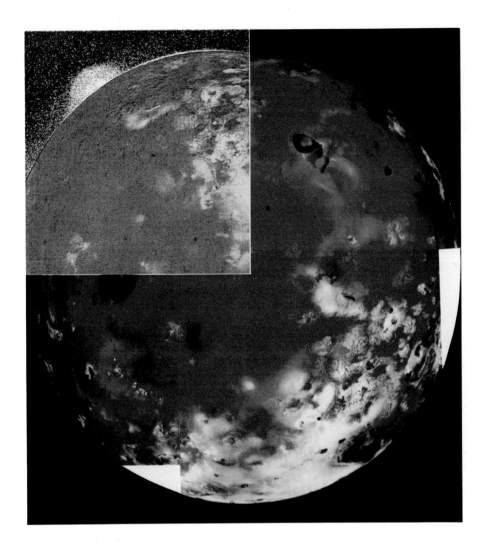

Plate III. Voyager 1 mosaic of Jupiter's satellite Io. Dark and mottled regions are associated with volcanos, and featureless areas with intervolcanic plains. (Inset) Active volcano on Io by Voyager 1, total height about 210 kilometers. The vent appears as a dark ring in the upper left region of the Io mosaic. (Courtesy of the National Aeronautics and Space Administration.)

Plate IV. The four Galilean satellites of Jupiter imaged by the Voyagers and brought to a relative scale. (Upper left) Io; (upper right) Europa; (lower left) Ganymede; (lower right) Callisto.(Courtesy of the National Aeronautics and Space Administration.)

agers' eyes, the outer edge of the ring nestles against the orbit of the tiny satellite XIV, also newly discovered and perhaps only 40 kilometers in diameter. The inner edges lie above Jupiter's clouds by 0.8 of the planet's radius. The brightest zone in the ring is about 800 kilometers across, surrounded by a fainter zone extending about 5200 kilometers and probably terminated by satellite XIV. The ring is not like the great rings of Saturn, which are made up of sizable pieces. The Voyager 2 was in the shadow of Jupiter when it imaged the ring in sunlight, as shown in Fig. 171. The ring in forward-scattered sunlight (looking in the general direction of the Sun) is brighter than in backscattered light, showing that the particles are extremely small. Possibly they extend down to the haze layer in Jupiter's high atmosphere and may actually be an extension of this haze layer. The ring of Jupiter cannot be a relict from ancient times. It may be continuously replenished from interplanetary dust (that is, cometary debris), from particles ejected by impacts on the great Galilean satellites, and/or possibly from volcanos on Io.

The orbits of the six inner satellites are nearly circular and arranged neatly in the plane of Jupiter's equator, with successive orbits averaging about 1.7 times the previous one, something like Bode's Law for the planetary orbits (Appendix 1). The eight outer satellites (Appendix 2) are tiny and their orbits are bunched in two sets of four, the first set near 12,000 kilometers from Jupiter in direct orbits and the second at nearly twice this distance, all in retrograde orbits. This opposed sense of revolution may protect the outer four from perturbational loss by the Sun's gravity, because they reach distances exceeding 0.2 astronomical units where the Sun's attraction can exceed Jupiter's by about a factor of two. The orbits are also relatively eccentric, up to 0.4, deviate up to more than 30° from Jupiter's equatorial plane, and of course vary considerably with time because of the Sun's attraction. Their stability over astronomically long periods of time is questionable. The outer satellites are darker than originally supposed and appear to be much like asteroids of the darker type found more in the outer part of the asteroid belt.

A remarkable fact about the three inner Galilean satellites— Io, Europa, and Ganymede—has long been known. They move in nearly perfect resonance, with periods of 1.77, 3.55, and 7.16 days respectively, or in ratios of 1, 2, and 4. Celestial mechanicians recognize this situation as a stable one, where the mutual perturbations hold the satellites permanently with these same relative periods even though tidal friction may change the periods. All of the inner satellites of Jupiter have been so affected and keep the same faces toward the planet. Just before the Voyager 1 approach to Jupiter, S. J. Peale, P. Cassen, and R. T. Reynolds showed that their mutual perturbations, mostly between Io and Europa, plus the effect of Jupiter's equatorial bulge, keep Io in an orbit with eccentricity of 0.0043 and Europa 0.011. The tidal heating effect varies as (eccentricity)^2 ×(density)^2 × (radius)^7 but inversely as (period)^5; so it turns out Io is heated about 20 times as much as Europa. Peale, Casson, and Reynolds predicted that the tidal heating in Io would amount to about ten times as much radioactive heating as in the Moon. Io, of course, may experience some additional radioactive heating. They stated that "consequences of a largely molten interior may be evident in pictures of Io's surface returned by Voyager 1." Their prophecy, consistent with the rapid pace of modern science, was immediately fulfilled (see Plates III and IV).

Eight volcanic plumes on Io were detected by Voyager 1, and six were active 4 months later as seen by Voyager 2. One volcano appeared to have died out between the missions and another was not observable because of the geometry. These plumes rise 70 to 280 kilometers about the surface, requiring ejection velocities as much as 1 kilometer per second. Forward scattering of the light suggests that the plumes contain fine particles, perhaps of sulfur dioxide, SO_2, which in an Io plume was detected as a gas by the infrared spectrometer on Voyager 1. Sodium and hydrogen have already been mentioned as present in Io's torus about Jupiter but no water vapor has been detected on or about Io. Present also are sulfur, and oxygen. Earth-based observations indicate the infrared absorption of SO_2 frost on Io, consistent with a vapor pressure of about 1 ten millionth bar at Io's equatorial temperature.

The volcanos on Io probably arise from a molten silicate interior having a small iron core. Io's mean density of 3.5 times that of water is slightly greater than the Moon's, supporting this conclusion. Beneath the visible crust lies an irregular silicate subcrust which, in a very few areas of small extent, protrudes as

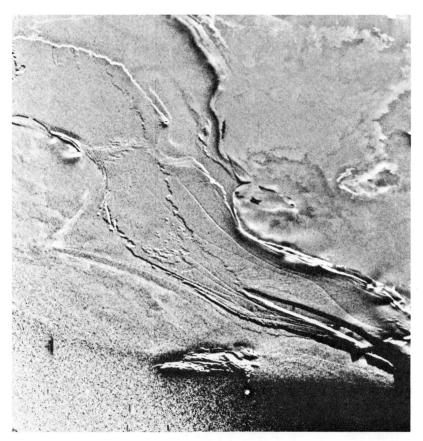

Fig. 172. Erosive flow in south polar region of Io. The flow markings are several hundred kilometers long. By Voyager 1. (Courtesy of the National Aeronautics and Space Administration.)

mountains not more than 10 kilometers in height. The temperature at modest depths may melt sulfur (120°C at 40 bar pressure), so that oceans of melted sulfur underlie the upper layer of solid sulfur mixed with SO_2. The sulfur, of course, is not as dense as the silicates, resulting in great flows of liquid sulfur that cover the surface and appear as huge plains with little relief (Fig. 172). The quantity of sulfur is unknown; the outer crust may possibly be thin and mixed with silicates. The polar regions are generally rougher than the equatorial regions, consistent with a lower temperature and less fluidity.

Much of the internal heat generated by tidal friction in Io is radiated away. But the currents in the molten interior of Io, as in the Earth, generate pockets of heat to produce volcanos, except that sulfur and SO_2 are shot out from Io instead of silicates and water as on Earth. The strong red, orange, yellow, brownish, black, and white hues on Io all support this concept of its upper layers. The colors are typical of sulfur and likely compounds involving oxygen, sodium, potassium, and hydrogen including nearly white areas of SO_2 frost. More than 300 likely vents have been mapped on Io. The largest is 250 kilometers across, lying in the equatorial region where more vents are seen. The more typical diameter is 40 kilometers.

No apparent impact craters with diameters greater than the resolving power, approximately 600 meters, are seen on Io. This indicates that resurfacing should amount to more than 0.1 millimeter per year, induced by plumes, flows, and surface erosion from volcanic action (Fig. 172). The present deposition rate from the observed plumes is estimated at 0.001 mm per year. Thus the new multicolored flows are probably younger than 1000 years. A few areas showed changes in the four months between the two Voyagers. Few features except the rare mountain tops are expected to exceed an age of one million years.

The present loss rate of sulfur and sodium to the Io torus is too small to seriously deplete Io's expected supply over the age of the solar system. Clearly the water has gone, literally boiled away ages ago, along with the other volatiles such as nitrogen and carbon, so prevalent on Venus and Earth.

When we look with the Voyagers at Europa, Ganymede, and Callisto we find totally different situations. Europa (Plate IV and Fig. 173) presents another striking surface, unique in the solar system. It is covered with a maze of tangled thin lines or streaks, indeed like the reputed "canals" of Mars. Some curve around for thousands of kilometers. Their widths are some 20 to 40 kilometers, probably cracks that are mostly filled. Europa is the smoothest body in the solar system, the tallest features rising only some 40 meters above the plains. Europa is like an orangish crystal ball, rather badly scratched. Few evidences of impact craters show on Europa, indicating practically complete healing of such wounds, almost instantly in astronomical time. The outer crust must be mostly ice to a depth of perhaps 100

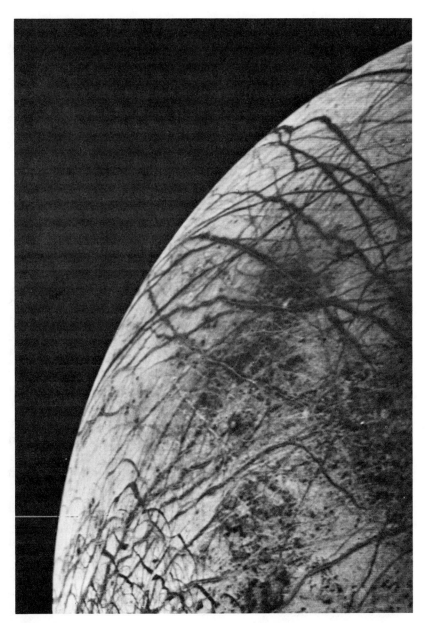

Fig. 173. Typical portion of Europa's surface by Voyager 2. The optical contrast is strongly enhanced. (Courtesy of the National Aeronautics and Space Administration.)

kilometers. The average surface temperaure is about $-150°C$ with the equatorial temperature rising somewhat higher at noon, consistent with a solid ice surface with small gas pressure. The infrared spectrum of Europa is dominated by water absorption and it, like Io, has a high albedo, characteristic of frost.

Because Europa receives about 5 percent of the tidal heating suffered by Io, its interior must be hot, but not hot enough to require special cooling mechanisms such as volcanos. Europa's interior composition may be essentially the same as Io's. The slightly lower mean density, 3.0 vs. 3.5 times that of water, could be accounted for by a 100-kilometer-thick ice or water crust. The internal heat could be conducted through a thin outer layer of ice, perhaps 10 kilometers thick; the huge array of cracks must be the consequence of strains and readjustments below the surface.

Far different is the surface of Ganymede, the largest and most massive of all the satellites. Its mean density, only 1.9 times that of water (Plate IV and Fig. 174), suggests a composition of about 50 percent water or ices by volume. Although Ganymede is a good reflector, approximately 40 percent, its surface is not as bright as Io and Europa, which have approximately 70 percent reflectance. Hence Ganymede is warmer, averaging about $-130°C$. Its rapid temperature changes during eclipse show that its upper surface layer is a poorer conductor than even the Moon's by nearly an order of magnitude. Furthermore, David Morrison and D. P. Cruikshank find that this insulating layer covers some 95 percent of the surface and is probably a thin layer of frost covering an ice or ice-dust-rock conducting layer.

Ganymede is anomolous in that its dark areas are saturated with craters, a few tens of kilometers in diameter, but the craters are extremely shallow. Here we have a record of ancient terrain, conceivably going back to the early bombardment period seen on the Moon, Mercury, and Mars. But how can this record be maintained on such a low-density body? In any case Ganymede shows a colossal ridge system thought to be an ancient impact basin on the unobserved part of the satellite.

Most intriguing are the bundles of grooved systems seen on Ganymede, some quite long and curving, like superhighways (Fig. 174). They cover much of its area, are unique to Gany-

Fig. 174. Close-up of Ganymede near latitude 15°, 1000 kilometers across. (Courtesy of the National Aeronautics and Space Administration.)

mede, and remain unexplained. The "geology" of Ganymede offers a number of challenging problems.

Proceeding now to Callisto, the outermost Galilean satellite, we find another unique body, the second largest in the solar system, and the least dense among the large satellites (Plate IV and Fig. 175). Furthermore, Callisto's surface on the far side from Jupiter is completely covered with craters, the most thoroughly peppered area yet observed (Fig. 175). On the near side, seen best by Voyager 1, is a huge multiringed structure with a bright central area some 300 kilometers across. Some 8 to 10 circular ridges surround the center to a distance of about 1500 kilome-

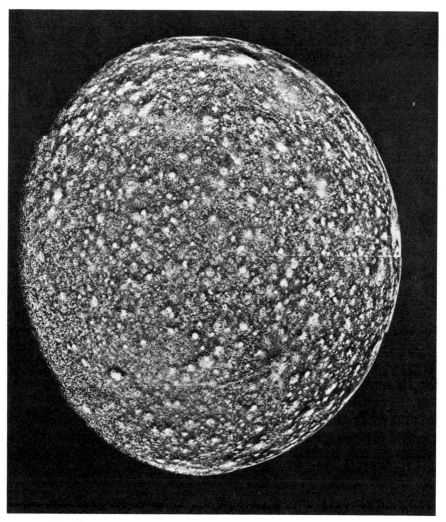

Fig. 175. High-contrast image of Callisto by Voyager 2, showing crater scars over entire surface. Note the great basin rings at upper right. (Courtesy of the National Aeronautics and Space Administration.)

ters, dwarfing the Caloris basin on Mercury and Procellarium on the Moon. The feature, however, shows no significant basin, implying mobility in the upper crust. The central area is much less cratered than most of Callisto's surface, suggesting that the impact was a relatively late event.

We may conclude that Callisto shows the scars of the great

bombardment of some 4 aeons ago. The large ringed feature and a second smaller one show that within the solar system, from Jupiter to Mercury, relatively large bodies contributed to the bombardment, many of them in the late stages. With a mean density of only 1.8 that of water, Callisto must contain more water than Ganymede, but still retains, with little topographic relief, the marks of ancient craters. Early in solar system history it had already developed a thick icy (permafrost?) crust which quickly filled the crater basins and still does. Callisto's low albedo, only 0.2, suggests an intermingling of dust. Because it reflects so little sunlight, it is the hottest of the Galilean satellites at its surface, about −120°C or warmer, still too cold to provide much water atmosphere, which is barely observed. Callisto's surface is also the best insulator among Jupiter's great satellites, with a dark outer crust that is highly porous and very thin, perhaps only millimeters thick.

The Galilean satellites, through the eyes of the Voyagers, are contributing vital knowledge about the evolution of the solar system, as we shall see later. Very early they made one major contribution to physical knowledge. When Olaus Roemer, a Danish astronomer (1644-1710), observed the eclipses of these satellites by Jupiter, he discovered that the time intervals between the eclipses were greater when the Earth was receding from Jupiter than when it was approaching. In 1675 he came to the conclusion that the apparent variations in the periods were caused by the *finite* velocity of light; previously light had been suspected of moving instantaneously. When the Earth is receding from Jupiter the light must travel a successively longer path between eclipses, while in approach the path is successively shortened. Roemer invented several of the most important instruments of positional astronomy. His great contributions, unfortunately, were little appreciated during his lifetime and only his proof of the finite velocity of light is much remembered today.

13

The Other Giant Systems— Saturn, Uranus, and Neptune

Saturn

Among the innumerable celestial objects that may be seen through a telescope, perhaps the most beautiful of all is the planet Saturn (Fig. 11). When viewed in the evening twilight while the sky is still bright, the yellow-gold ball and its unbelievable rings shimmer in a brilliant blue medium, more like a rare work of art than a natural phenomenon. Lightly shaded surface bands, more uniform than those of Jupiter, parallel the great rings; only occasionally can one distinguish detailed markings that will reveal the rapid turning of the great globe. The central brilliance fades away toward the hazy limb of the planet's disk, and the rings at their borders appear to dissolve into the sky.

Where Saturn's rings cross the disk a hazy dark band outlines their innermost edge (Fig. 176). This "crape ring" is most readily discerned by its faint shadow on the planet's disk. The outer rings also cast shadows on Saturn, which in turn completely

1915   1921

Fig. 176. Saturn's rings disappear when seen edge-on. At left the rings are near their maximum opening in 1915. At right, the rings are edge-on, in 1921. See Fig. 10. (Photographs by E. C. Slipher, Lowell Observatory.)

eclipses large sections of the rings. The polar regions of the planet, perpendicular to the plane of the rings, are darker than the other limbs of the disk, and, when seen under good observing conditions, present a slightly greenish appearance. Three major bright divisions of the rings are easily detected, the brilliant middle ring (B), the fainter outer ring (A), and the barely luminous crape (C). The two outer rings are broken by narrow dark gaps, similar to Cassini's division, which separates rings A and B (named for G. D. Cassini, 1625–1712).

Astronomers were awed by the vastly changed appearance of the rings when NASA's Pioneers passed Saturn in 1979 and gave us for the first time a view from beyond Saturn, with sunlight shining *through* the rings toward us. Awe changed to amazed excitement when in November 1980 Voyager 1 sent back the now classic picture of Fig. 177. Backlighted by the Sun, the familiar smooth rings are transformed into possibly a thousand concentric narrow ringlets, not unlike the grooves in a phonograph record. The great Cassini division is barely identifiable, containing four conspicuous bright rings. Nevertheless, as expected, the division allows sunlight to brighten Saturn's disk between the shadows of the A and B rings. A new era in Saturnian research began with this picture.

Saturn's rings lie precisely in the plane of the planet's equator, inclined some 28° to the plane of the Earth's orbit. Because

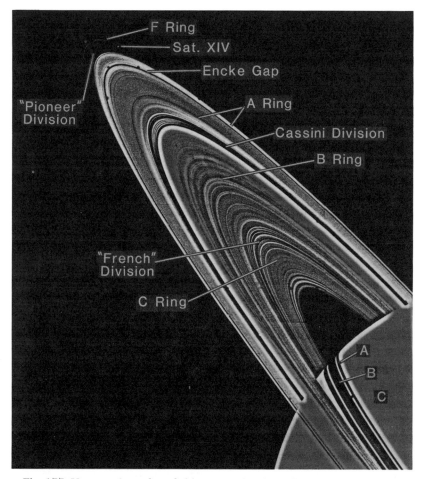

Fig. 177. Voyager 1 produced this composite view of Saturn's complex ring system with the Sun shining through the rings. The major rings are almost unrecognizable except by their shadows on the disk of Saturn. (Courtesy of the Jet Propulsion Laboratory for the National Aeronautics and Space Administration.)

the plane of the rings remains fixed as Saturn moves around the Sun, during a revolution we can see the rings from above (north), from below (south), and twice edge-on (Fig. 178). When the rings are tipped at the greatest angle for terrestrial observation they reflect nearly twice as much sunlight as Saturn itself, but when the rings are edge-on they practically disappear for a short interval. The thickness of the rings is therefore exceedingly small. The rings were last seen edge-on from the Earth on October 27, 1979, and on March 12 and July 23, 1980. In this century they will again be edge-on May 21 and August 11, 1995, and February 11, 1996.

The rings are composed of individual fragments of matter, each moving in its own orbit about Saturn according to Newton's law of gravitation. The irrefutable demonstration of this fact is made with the spectrograph. We have seen in Chapter 11 how the spectrograph can be used for identifying a gas by the spectral lines, or wavelengths, that are subtracted from the light when it passes through the gas. If the gas is approaching us, or we are approaching the gas, the waves appear to be closer together because of the relative approach, with the result that all the wavelengths are measured shorter than before. Since wavelength decreases toward the violet end of the spectrum, all the dark lines of missing wavelengths are shifted toward the violet. The amount of the shift is proportional to the velocity of approach. When we are receding from the source of light, the waves are apparently spaced farther apart, and the dark lines are displaced toward the red end of the spectrum. This phe-

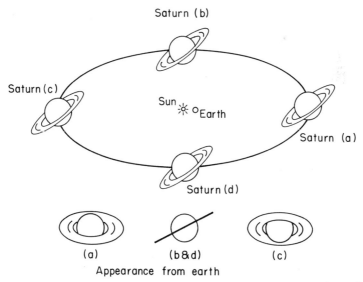

Fig. 178. Changing views of Saturn's rings as seen from the Earth. Positions *a* and *b* correspond to the photographs in Fig. 176.

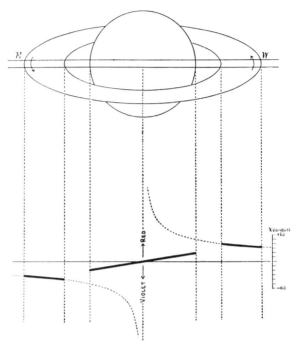

Fig. 179. Rotation of Saturn and its rings. Diagram to explain the spectrum. Note the tilt of the lines, showing that the outer edges of the rings move more slowly than the inner. The scale at the lower right represents radial velocity in kilometers per second. (Drawings by J. E. Keeler, the Yerkes Observatory.)

nomenon is known as the Doppler-Fizeau effect. In sound the analogous Doppler effect is exhibited by the drop in pitch of the bell of a train as it passes.

The tilts of the lines in Fig. 179 display the shifts in the wavelengths to demonstrate the rotation of Saturn and motions in its rings. Note the reverse tilt of the lines in the rings as compared with the disk. At every point along the rings the measured velocity agrees exactly with that of a corresponding satellite if it were moving in a circular orbit, more slowly with increasing distance from the center. Kepler's laws of motion are obeyed precisely. Were the rings solid, the outer edges would move more rapidly than the inner. This demonstration of the discontinuous structure of Saturn's rings was performed in 1895 by J. E. Keeler (1857–1900) at the Allegheny Observatory. The velocity difference across the A and B rings is about 6 kilometers per second.

On rare occasions the rings will cross the line of sight to a star, occasionally the satellites pass behind the rings, and rarely we can see some of the satellites when shadowed by the rings. Stars disappear when obscured by the densest part of ring B, flash up to nearly normal brightness at Cassini's division, and show irregular fluctuations in ring A, now thoroughly explained by the many ringlets shown in Fig. 177. A small amount of light diffuses through even the densest part of ring B to give a very faint illumination of a satellite in the darkest shadow. The rings are slightly redder than sunlight and the thickest part of ring B has a high albedo of 0.70 in visual light and 0.57 in blue light. This observation is consistent with the rings being covered with ice or frost, as the spectrograph confirms.

The rings brighten up remarkably when observed almost at precise opposition from the Sun. Fred A. Franklin finds that in the last degree there is an increase of more than 30 percent. In 1887 H. von Seeliger (1849–1924) suggested that the many small particles in the rings shadow each other so that only at opposition can we see the surface of the particles directly without appreciable effects of shadowing. The phenomenon is similar to the *heiligenschein,* the halo that appears about one's head in long shadows on dewy grass in the early morning (see Fig. 180). In matted vegetation, looking near the line of sight from the Sun one can see the light reflected directly from all surfaces. But at appreciable angles to the line of sight, the light entering the vegetation has to find a new route out to the eye and consequently is dimmed. To an observer in a high-altitude aircraft the shadow of an aircraft on the ground appears as a bright spot; over water it turns into a dark circle.

The bizarre appearance of Saturn's rings when backlighted by the Sun (as in Fig. 177) involves yet another optical effect. Sizeable particles much larger than the wavelength of light (that is, much larger than half a micron) cast shadows and scatter light, mostly back toward the light source. Tiny particles, comparable to or smaller than the wavelength of the light, tend to scatter light forward instead of casting strong shadows (analogy: headlight glare on a dusty windshield). From the Earth we see the light scattered backward from larger particles in Sat-

Fig. 180. The *heiligenschein* effect. Photographed by William Sinton from the dome of the 200-inch reflector at Palomar. Look at the shadow of the dome and note how the brightness of the forest drops rapidly from the edge of the shadow.

urn's rings. But Pioneer and Voyager, beyond Saturn, have also imaged the light scattered forward by the tiny particles, which become conspicuous when the Sun is behind them, though they do not contribute much to the mass of the rings nor to shadows on the disk of Saturn. Thus, some of the dark rings in Fig. 177, such as those in Cassini's division, indicate a true absence of particles, while others indicate a high density of large particles that block the sunlight. The C ring, which appears as a hazy shadow from Earth, becomes bright when backlighted because of its many tiny components. Extensive measurement and research will be required just to describe fully these complex ring structures.

Pioneer and Voyager have confirmed the E ring, faint and extensive beyond the bright rings, and the Encke gap, breaking the outer edge of the A ring. They have clearly delineated the "French" division between the B and C rings and defined a very faint inner D ring. The "Pioneer" division beyond the A ring is bounded by the newly discovered F ring, less than 150 kilometers wide. A faint G ring may lie between. The major dark divisions (not all officially named) are associated with ring particles

which would have to move with periods that are simple fractions of the periods of the satellites Mimas, Enceladus, and
Tethys, as listed in Table 5. A complete theory of how these resonances affect satellite orbits is yet to be worked out.

The placing of the gaps is also affected by the attraction of
Saturn's equatorial bulge and by the masses of the rings, particularly the massive B ring, whose pieces may add up to a broken
(or unformed) satellite several hundred kilometers in diameter.
This is true if the typical particles are roughly 1 to 15 meters in
diameter, with some larger and smaller, consistent with the
radar reflections first obtained by R. Goldstein and G. A.
Morris.

The myriad of Voyager ringlets within the great rings present a great proving ground for theories. Particles in the rings
cannot coalesce into sizeable satellites because Saturn's tides will
tear them apart (see p. 282). At the same time, the inner ones

TABLE 5. *Features of Saturn's ring system.*

Feature	R (*Saturn's radius* $= 1.0$)
C ring	1.22–1.50
"French" division ($\frac{1}{3}$ period Mimas)	1.50–1.53 (1.49)
B ring	1.53–1.96
Cassini division ($\frac{1}{2}$ period Mimas) ($\frac{1}{3}$ period Enceladus) ($\frac{1}{4}$ period Tethys)	1.96–2.03 (1.94)
Encke Gap ($\frac{3}{5}$ period Mimas)	2.21[a] (2.21)
A ring	2.03–2.27
Pioneer division	2.27–2.33
F ring ($\frac{2}{3}$ period Mimas) ($\frac{1}{3}$ period Tethys)	2.33[a] (2.37)
E ring	to 5–6
Satellite XI ($\frac{1}{2}$ period Enceladus)	2.53 (2.51)

a. Narrow.

Fig. 181. The "braided" F ring as imaged by Voyager 1. The knots may be mini-satellites or mass concentrations. (Courtesy of the Jet Propulsion Laboratory for the National Aeronautics and Space Administration.)

must continually overtake and pass the outer ones in their gravitational motion. Do they minimize collisions by lining up in narrow rings? Do larger unidentified masses or satellites control the ringlets? Are higher-order satellite resonances responsible? How can ringlets form in the dark divisions and why are a few measurably eccentric? And what about the "braided" F ring (Fig. 181), where three discrete rings or streams appear to be interwoven?

Another Voyager puzzle involves its verification of occasionally observed radial "spokes" in the main rings (Fig. 182). Their evanescent changes suggest tiny charged particles streaming in Saturn's magnetic field and turning with it. Particles constrained by gravitational forces alone could not produce such radial structures.

The Voyagers are expanding Saturn's system by adding several more satellites, but the orbit of Janus (X) is not confirmed. The new satellites are small but somewhat larger than Jupiter's tiny ones. Two move on either side of the F ring and appear as broken rocks—"herding" the ring particles by gravity? A new satellite created a sensation because it apparently shares another's orbit, both moving with nearly half the period of Enceladus. Dione gravitationally controls a small satellite moving in Dione's own orbit. Close-up pictures of the inner named satellites except Enceladus reveal the usual pock marks of battered surfaces and evidence of deep cracks and valleys. Mimas (Fig. 183) shows a huge crater pointed forever at Saturn. Infrared measures disclose the presence of water ice on Enceladus, Tethys, Dione, and Rhea. All of the larger ones whose masses can be approximated appear to have densities between 1.0 and 1.3 that of water, except for great Titan, whose density is a little more than twice that of water. Voyager's close approach to Titan has reduced earlier estimates of its size, leaving Gany-

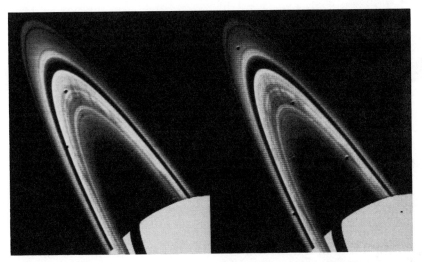

Fig. 182. Dark radial "spokes" in Saturn's B ring, imaged 15 minutes apart by Voyager 1. Note the displacement of the spokes, indicating a rotation period about equal to that of Saturn itself. The small dots are produced by the camera. (Courtesy of the Jet Propulsion Laboratory for the National Aeronautics and Space Administration.)

Fig. 183. Saturn's satellite Mimas imaged by Voyager 1, bearing age-old scars of impacts, probably mostly by comets. The impact that produced the crater greater than 100 kilometers in diameter must have been a near disaster for Mimas. (Courtesy of the Jet Propulsion Laboratory for the National Aeronautics and Space Administration.)

mede of Jupiter as the largest and most massive satellite of the solar system.

Titan, however, remains unique as the only satellite with an extensive atmosphere. In 1943–44 Gerard P. Kuiper photographed methane in its spectrum, about half as much as in the atmospheres of Jupiter and Saturn. Voyager startled us by revealing that the atmosphere of Titan is huge, perhaps much more than 10 times the Earth's in mass per unit area, and consists primarily of nitrogen. The actual surface remains unobserved. Speculation ranges from the possibility of liquid-nitrogen rain and oceans on Titan, to the much more remote possibility that the surface may be warm and even water covered. Polar haze layers are evident in the cold high atmosphere, possibly particles of methane or of complex hydrocarbon molecules. The dissociation of methane, CH_4, from Titan's outer atmosphere may account for the extensive hydrogen cloud around Titan. The cloud is widely spread within and beyond the orbit, observed both by Pioneer and by Voyager.

Saturn's ninth and most remote satellite, Phoebe, revolves in a retrograde orbit 8 million miles in radius. When discovered by W. H. Pickering in 1898, it was the only satellite whose motion was known to be retrograde with respect to its primary, but its claim to fame was later shared by other retrograde satellites.

Iapetus, the sixth satellite, is unique in that its leading hemisphere is extremely dark, reflecting about 5 percent of the incident light, while the following hemisphere reflects six times as much. Iapetus, like the Moon and many other satellites, rotates with the same side toward its primary. B. T. Soifer and associates found, and Voyager confirmed, that the bright side is covered mostly with ice while the dark side has less than 5 percent ice. The dark material is not like lunar dust, but a dark asteroidal material. Has the leading face of Iapetus added more dust by encounters, or have the encounters with particles caused it to lose its icy face?

Saturn itself is unique in the solar system because of its low mean density, only 0.7 that water. Like Jupiter, Saturn rotates very rapidly, in just over 10 hours, and is correspondingly flattened. The clouds on Saturn are much less conspicuous than on Jupiter. Large-scale disturbances occasionally appear, such as the great white spot of 1933 which disappeared a year later. Voyager detected various cloud patterns and a "red spot" about the size of the Earth.

The temperature at the cloud tops is some 30°C cooler than on Jupiter, -178 to -173°C, but the hydrogen–helium atmosphere extends above the clouds, much as for Jupiter, at a pressure less than one Earth atmosphere. Also there is haze above the clouds, which may be composed of particles, possibly ice, descending from the rings, or of methane.

The Pioneer Saturn found that Saturn's magnetic pole is centered almost exactly along the pole of rotation in contrast to the skewness observed for the Earth and Jupiter. Its strength at the equator is about 0.7 that of the Earth but the intrinsic field is far greater because of Saturn's size. Theories for the source of the magnetism place the field-producing currents much deeper than for Jupiter but they give the same overall polarity, opposite to that of the Earth. The rings absorb the energetic ions and electrons of Saturn's magnetosphere, inferred from an abrupt cessation of counts by Pioneer Saturn as it passed under the edge of the rings. The satellites also cut swaths through the magnetosphere. The rings thus literally eat out the heart of Saturn's Van Allen belts, leaving the magnetosphere nearly impotent as a radio-noise source. The alignment of the poles still further reduces radio outbursts.

Most interesting from the evolutionary viewpoint, Saturn radiates from internal sources about 2.5 times the energy it receives from the Sun. Similar to the case of Jupiter, this energy far exceeds the amount expected from radioactivity. There must still be some contraction or settling of material to the center, but not enough to be observationally perceptible for aeons.

Uranus and Neptune

These two planets are nearly identical twins, giants in the outer regions of the solar system. Their diameters are about four times that of the Earth; Uranus is the larger by perhaps 5 percent, although the measures are somewhat uncertain because of the hazy edges of the disks. Neptune, however, is the more massive, comprising 17.3 Earth masses while Uranus has 14.5.

Although surface markings are difficult to observe on Neptune (Fig. 184), perhaps because of the great distance, and only faint belts have been seen on Uranus, the planets are certainly enveloped with atmospheres resembling those of Jupiter and Saturn. The albedos are high and the spectra show methane absorptions similar to those for Jupiter and Saturn, but much intensified. Observe the sequence of spectra in Fig. 164. The absorptions of yellow and red light by methane vapor are so enormous for Uranus and Neptune that the planets appear greenish in color when observed directly; the color is more pronounced for Neptune. The spectrograms evidence no certain trace of ammonia, but hydrogen is present.

A lack of gaseous ammonia and an abundance of methane are readily explained as resulting from the vast distances of these planets from the sun and the corresponding diminution in the amount of solar heat received at their surfaces. The sunlight is indeed so weak for Uranus that its surface temperature is about $-215°C$, while Neptune is about 2 degrees colder. At its much greater solar distance Neptune should theoretically be some 12 degrees colder. Apparently Neptune has an internal heat source about equal to the solar input while Uranus does not. Explanations for the existence or absence of internal heat

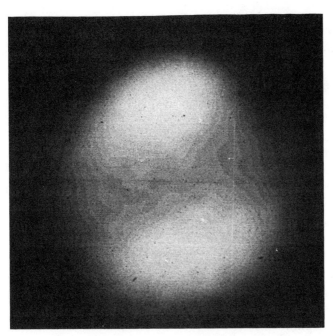

Fig. 184. Neptune imaged in the dark infrared band of methane. The bright regions represent high-level methane crystal clouds where the methane layer above is thinner. Taken with the Catalina Observatory 154-cm reflector of the University of Arizona. (Courtesy of Bradford A. Smith.)

sources for the giant planets remain in the gray area of theory. No compelling basic causes are evident except probably continued contractual evolution. But why for Neptune and not for Uranus?

The increasing strength of methane absorptions and the weakening of the ammonia absorptions as we progress from Jupiter to Neptune (again Fig. 164) certainly arise from the decrease in temperature. The vapor pressure of ammonia decreases very rapidly with decrease in temperature; hence precious little ammonia gas can remain in Neptune's atmosphere and, indeed, methane will freeze there at high enough pressures. In Fig. 184 we see an infrared photograph of Neptune taken in the weak light from the center of the methane absorption band. Because the equatorial regions are darker than regions at higher latitudes, we are seeing deeper into the methane atmosphere at the equator. Thus the clouds on Neptune are lower near the equator than at the poles and probably consist of methane particles, as do the clouds on Uranus.

The internal heat, even for Neptune, is only a few times that of the Earth, so that neither Uranus or Neptune has much of a problem in carrying heat to the surface for radiation into space. Consequently their clouds systems are extremely weak compared to those of Saturn or Jupiter and hardly observable until spacecraft or the space telescope can observe them. Neptune, in fact, may have haze that blankets any clouds below. It appears to show less limb darkening than Uranus.

The periods of rotation remain in question, determinations having been made from spectrographic line shifts, from photometry, and from calculations based on difficult measures of oblateness. Uranus probably spins in approximately 16 hours about an axis tilted by 98° to its orbital plane. The fact that the five satellites move so nearly in a plane is almost conclusive evidence that the planet rotates in the same plane. Neptune's period is probably about the same, in the range 11 to 19 hours. Neither planet shows much evidence of a strong magnetic field, although W. L. Brown has detected radio pulses near 600 meters wavelength from Uranus.

The Rings of Uranus

The discovery of the rings of Uranus illustrates the happy surprises that occasionally drop out of otherwise rather prosaic projects. In this case the goal was to measure the diameter of Uranus with greater accuracy and to learn more about its upper atmosphere by observing the occultation of a ninth magnitude star as it passed behind the planet. On March 10, 1977, J. L. Elliot, D. Dunham, and D. Mink flew the Kuiper Airborne Observatory, KAO, from Perth in Australia to observe the occultation under ideal conditions at 12.5 kilometers altitude. The occultation was expected to be seen south of Perth, and so the KAO flew to 52° south latitude over the South Pacific. The telescope was fortunately centered on Uranus and the photometric equipment set in action more than three quarters of an hour before the expected beginning of the occultation. Almost immediately the photometer registered a quick deep dip, or recorder spike, in the star's brightness—a new satellite of Uranus? This

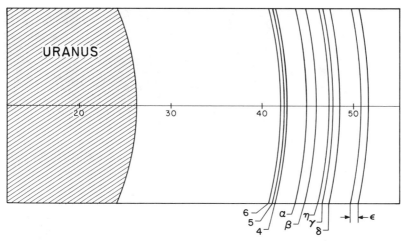

Fig. 185. Schematic projection of the Uranus ring system on the equatorial plane. Of the nine rings, only ε shows appreciable width.

happened five times before the occultation of the planet began, some 5 minutes late. Excitement and uncertainty prevailed until finally a reduction of the 12 meter-long data tape revealed that the 5 spikes before the occultation matched 5 more after occultation at times equivalent to corresponding distances from Uranus in the plane of its satellites. Four rings (α, β, γ, δ) in nearly circular orbits and one (ε) in a slightly eccentric orbit had been discovered! Confirmation came from observations by four other groups.

Nine such rings have now been identified, all between 41,000 and 52,000 kilometers from the center of Uranus, or above its surface by less than its radius (Fig. 185). Seven are extremely narrow, about 0.5 to 3 kilometers across and very opaque. Ring β is wider, approximately 12 kilometers, and ε, the most distant, is quite different from the rest. Its width, 20 to 100 kilometers, varies linearly with its distance from Uranus, but, judging by its transmission of light (10 to 60 percent), the integrated area of the obscuring particles is nearly constant around the ring.

Apparently this eccentric ε ring is precessing as a whole (as a solid body would) around Uranus with a period of some 8 or 9 months, the precession being caused by the equatorial bulge of Uranus. Particles in the ε ring have a period of about 0.354 days whereas the period of the inner satellite Miranda is almost ex-

Fig. 186. Uranus (mostly obscured by diaphragm) and all five satellites. They are (top to bottom) Oberon, Umbriel, Miranda (very faint just below Uranus), Ariel, and Titania. (Official U.S. Naval Observatory photograph, by R. L. Walker.)

actly 4 times this value, 1.414 days. The ϵ ring's period is also nearly in resonance with $\frac{1}{7}$ the period of Ariel, the second satellite. The rings 5, α, γ, and ϵ all may be controlled by resonances of these two inner satellites. Resonance control for the other rings is not so well established. Quite possibly smaller undiscovered satellites are involved.

The five satellites of Uranus are all small and cannot be well observed as yet (Fig. 186; see also Appendix 3). They all move very closely in the same plane, tilted 98° to Uranus' orbit, probably defining the planet's equator better than direct observations. Their masses are only roughly estimated from their mutual perturbations and their radii are so uncertain as to make density determinations meaningless.

Neptune's Triton is one of the largest satellites in the solar system, considerably larger than the Moon, with about 1.9 its

Fig. 187. Gerard P. Kuiper (1906–1973), a major contributor to solar-system astronomy, who discovered the satellites Miranda of Uranus and Nereid of Neptune.

mass. Thus the density, poorly determined, may be in the range 1.5 to 3.0 times that of water. Although Triton has an extremely tenuous methane atmosphere, its surface is rocky, not icy. Nereid is small and little is known about it (Fig. 187). The two satellites, however, show a remarkable disparity in their orbits. Whereas Triton moves in a retrograde motion about Neptune at an angle of some 28° to the plane of the planet's equator at a distance of some 356,000 kilometers, Nereid moves in direct motion at a much greater distance of some 5,570,000 kilometers. Thus, Triton is the only inner satellite in the solar system with a retrograde orbit. As such, its tidal friction with Neptune may place it in jeopardy. Thomas McCord suggests that Triton might spiral to destruction in 10 to 100 million years.

The giant planets, in spite of, or because of, their huge dimensions, provide no possible abode for life of the kind we know, nor do their satellites, except possibly for Saturn's Titan.

14

Comparisons among Planets and Satellites

Observations of actual stellar formation in the dark gas and dust clouds of interstellar space now convince us that our Sun and its planetary system originated by the collapse of part of such a cloud, not unlike an hypothesis first made by Immanuel Kant in 1755. Probably our system was one of many in a stellar incubator, born with giant stars and perhaps nudged by supernovae. The bodies in our system were derived from the same source material as was the Sun. We know the composition of the Sun and the meteorites and much about the composition of the Earth and of the Moon's surface. We also know a great deal about masses, densities, and other physical characteristics of most of the planets and several satellites. Combining all this knowledge with known physical laws, we are now in position to speak knowingly about the composition and internal structure of several planets and satellites.

By analogy we may say that we have deciphered a part of the grand recipe for making planets and satellites. We know the basic ingredients. We know the proportions of the ingredients distributed among the various bodies and much about how they are layered within these bodies. This is *comparative planetology,* the subject of this chapter. The more subtle nuances of the recipe such as stirring, separating, baking (temperatures and times), decanting, coagulating, cooling, peppering, and general timing remain relatively vague and speculative, the subject of the next chapter on the evolution of the solar system.

We cannot directly explore even the Earth to more than about 0.1 to 0.2 percent of its radius. We are forced to combine data and theory to determine the internal structure of a planet or satellite. We can use three classes of information: (1) the gross physical properties, from direct observations and inferred from gravitational action, (2) the possible or likely materials available from the solar abundances of the elements, and (3) the chemical and physical properties of these materials, particularly at very high pressures and high temperatures.

We now know the shapes, sizes, masses, mean densities, and surface characteristics for eight planets and at least six large satellites. Earlier we saw how seismometers have measured earthquake waves to give us remarkable inside information about the Earth. Similarly, moonquakes gave us a surprise about the Moon's interior. Unfortunately we have no seismic information about any other celestial body. Only one insensitive seismograph operated on Mars, giving no record of any marsquake. At least we know that *large* quakes are very infrequent on Mars.

The polar flattening of rapidly rotating planets tells us something important about their interiors. The equatorial bulge attracts nearby satellites or passing spacecraft differently than it attracts bodies at great distances. We saw how the Sun's and Moon's attraction for the Earth's equatorial bulge causes the precession of the equinoxes, the spinning-top effect. These gravitational effects from an equatorial bulge measure the density of the upper layers, to be compared with the mean density of the whole planet. This adds up to a knowledge of the *moment of inertia* for the body, that is, how much force is required to set it spinning or to slow it down. We know this important quantity for the Earth, Moon, Mars, Jupiter, and Saturn, but poorly for Uranus and Neptune and not at all for Mercury and Venus because they rotate so slowly.

Another bit of information comes from magnetic fields or their absence. Strong fields such as those on the Earth, Jupiter, and Saturn tell of hot internal convection currents. We have other clues about interiors from surface features such as volcanos and craters, old or new, and particularly from the general topography. For example, the nature of the great Tarsus uplift on Mars and the planet's general surface elevations, along with records of cratering, indicate a limited or perhaps single internal circulation event, that is, limited plate tectonics on Mars. In contrast, the minimal elevation profile on Europa tells another story of rapid crustal readjustment by ice flow in the upper layers. Atmospheres, their circulation patterns, and evidences of internal planetary heat sources provide vital information.

Gravity measures over the Earth mirror the thickness and density of the upper crustal layers. For the Moon, orbiting spacecraft give much the same information. An appreciable amount is being learned about Mars from orbiters, and eventually we may expect the space program to provide such data for all the planets and major satellites, although today this knowledge is limited.

Radioactivity can melt planetary interiors and greatly affect their structures. Present-day interiors therefore reflect their thermal histories, which must be taken into account in the calculations. Lack of current evidence for heating is also a diagnostic tool. Fortunately we know the time scale for the solar system, 4.6 aeons, which is immensely valuable for calculating the heat contributed by the known radioactive elements. But we must admit the possibility of unobserved short-lived contributors in the early days.

Let us look now to the materials available for making planets and satellites and some of their properties. The Sun is rather well mixed except perhaps at the very center. Solar modeling gives the best values for the hydrogen and helium abundances, while spectroscopy of the deep atmosphere is reliable for the relative abundances of the other elements. For our needs we divide the elements into three classes: (1) Gases: hydrogen, helium, and the noble gases; (2) Ices: carbon, nitrogen, and oxygen plus some hydrogen; (3) earth or rock: silicon, iron, and all the other elements. The reason for these classes is, of course, temperature. At very high temperature as in the Sun, everything is gaseous. In low to moderate gravities at temperatures between 700°C and room temperature, the rocky material will be solid, combined with an appreciable fraction of oxygen, and the molecules of carbon, nitrogen, oxygen, and hydrogen will be gaseous. At very low temperatures all the materials except hydrogen, helium, and the noble gases will be solids. These divisions of our basic materials are fundamental to understanding planetary evolution.

It turns out that only a very small fraction of the basic materials are available for making earthlike planets. The rocky component is only 0.4 to 0.5 percent of the total, or $\frac{1}{200}$ to $\frac{1}{250}$. The icy fraction is somewhat larger, 1.8 percent, giving us about 2.2 percent of the total or about $\frac{1}{50}$ for making comets and planets and satellites compounded from the combination of icy and rocky materials. We could profitably break the rocky division into many subdivisions of minerals that play important roles in the formation of the terrestrial planets, the asteroids, and meteorites. Suffice it to say that iron constitutes about a third of the rocky mass, with silicon, magnesium, and sulfur adding another third. If we assume that atoms of these elements typically combine with two oxygen atoms to form minerals, we raise our potential rocky fraction of the total to about 0.7 percent or $\frac{1}{140}$ part. About 10 percent of this is sulfur dioxide, which, as we have seen for Io, sometimes acts as a volatile substance rather than a rock.

The physical laws for very hot gases compressed in stars are better known than those for lower temperatures in planets because the laws are simpler for gases than liquids or solids. The Earth's gravity exerts pressures up to millions of atmospheres at its core, beyond laboratory capabilities at temperatures of several thousand degrees. Various minerals compress differently, adding to the difficulty of calculating internal properties of the Earth. For the giant planets the pressures and temperatures are even greater. The central pressure at the earthy core of Jupiter is believed to be some 300 million atmospheres at approximately 30,000°C. The theory for the compression of hydrogen (H_2), the major component in most of the body of Jupiter, is well known perhaps to a depth of 25,000 kilometers. Under greater pressure, hydrogen changes its state to liquid, called metallic because the electrons are freed, as in metals, to conduct electric currents freely. The theory for metallic hydrogen, helium, and other materials under great pressures and temperatures is constantly being improved so that successive "model" interiors for the planets show less and less variation with time.

The Planets

The terrestrial planets contain only rocky materials, as illustrated in Fig. 188. Mercury shows a large proportion of iron so that the contribution of oxygen combined in minerals is relatively small. The Earth and Venus appear to contain combined oxygen in considerable proportions, as does Mars. The composition trend outward from the Sun suggests falling temperatures, or the loss of a large fraction of the materials lighter than iron for Mercury near the Sun.

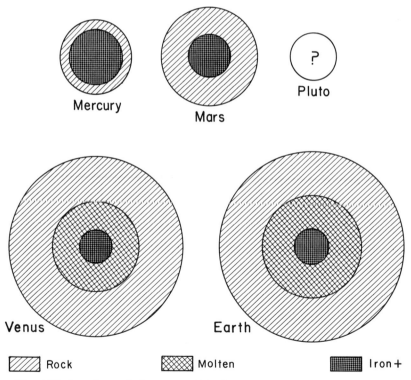

Fig. 188. Interiors of the terrestrial planets. Venus is modeled like the Earth. The Mercury model follows Norman Ness, and the Mars model an average of several.

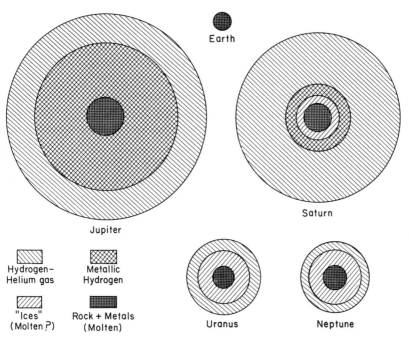

Fig. 189. Interiors of the giant planets. The core of Jupiter is little understood, here indicated as a mixture of "icy" and "rocky" material at extremely
high temperatures. Jupiter and Saturn follow models by M. Podolak and
A. G. W. Cameron; Uranus and Neptune after W. B. Hubbard and J. J. Mac-
Farlane.

For the giant planets the picture in Fig. 189 is entirely different. Great Jupiter is mostly gaseous, being close to the Sun in
composition. Surprisingly, Jupiter is almost the maximum size
for a "cool" planet (Fig. 190), one having no strong energy
source. Putting more mass into Jupiter would increase the gravity to reduce the diameter. Somewhat less mass would not
greatly expand the planet unless more heat were introduced.
Note that Saturn is nearly as large as Jupiter with less than one-
third its mass and about one-half its mean density.

The next step outward from the Sun, Saturn to Uranus and
Neptune, is striking. These outer giants are much less massive,
much denser, and much smaller. The remarkable fact, however, is that the absolute content of ices plus rocky material is
roughly the same for all four planets. Present-day theory is still
inadequate to give a very accurate central composition for Jupiter, but its maximum content of elements heavier than helium
is rarely placed above 30 to 50 Earth masses by any investigator.

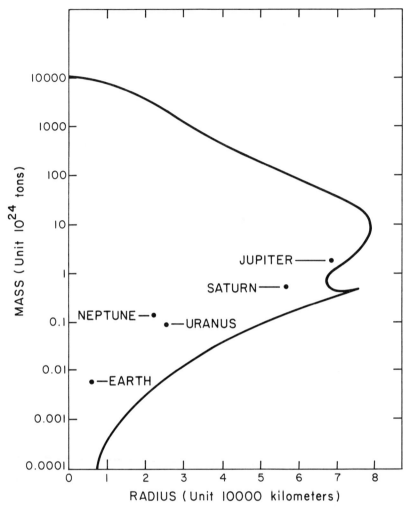

Fig. 190. Radii and masses of cold hydrogen spheres compared to actual values for the Earth and giant planets. (After W. C. DeMarcus.)

In more recent studies the estimates lie more in the range of 15 to 20, very comparable to the values for the outer three giants.

As we saw earlier, the solar proportion of the ices plus rocky materials—the "cometary" fraction—is only about 2 percent. Thus the original mass from which the giant planets were fractionated is more or less one Jupiter mass each. Neptune, however, is about 94 percent of cometary composition and Uranus about 89 percent. The broad scenario of giant planet origin is evident. Uranus and Neptune must have aggregated from cometary-type material, a little gas being added. Jupiter and Saturn concentrated the original solar-mix of elements, to which some cometary material was collected. It is difficult to accept the alternative theory in which Uranus and Neptune had once been comparable to Jupiter or Saturn in mass and then lost their gases. At such great solar distances no likely loss mechanism is evident.

On the other hand the rocky content of the terrestrial planets suggests two diametrically opposed scenarios of aggregation. Their individual share of the original solar mix of elements is perhaps one-quarter, compared to the individual share for the giant planets. They may have been aggregated from purely rocky materials at more or less room temperatures. Conversely, they may be the remaining cores of large protoplanets of solar composition. These opposed possibilities will be discussed in the next chapter.

The Satellites

Turning now to the major satellites as in Fig. 191, we see a sequence of compositions involving only cometary type materials, ranging from purely rocky to icy-plus-rocky materials. The Moon somewhat simulates the upper mantle of the Earth, iron probably being relatively deficient. The complete lack of water and other volatile elements on the Moon suggests considerable heating and loss of volatiles. Current opinion favors a relatively cold Moon in its early stages. Radioactivity then added internal heat while the upper layers were heated by accretional bombardment coupled possibly with electromagnetic heating from the expected solar gale when the Sun was very young. Clearly the Moon was expanding at the time of the great bombardment about 4 aeons ago, attested to by the huge number of radial cracks focusing on Mare Imbrium.

In contrast, Mercury may have been contracting at the time when the great Caloris basin probably was formed. The difference may well lie in the percentage of radioactive elements. They tend chemically to separate out with Earth-mantle minerals, not falling to the center with the iron. Hence they may well be less abundant in Mercury than in the Earth.

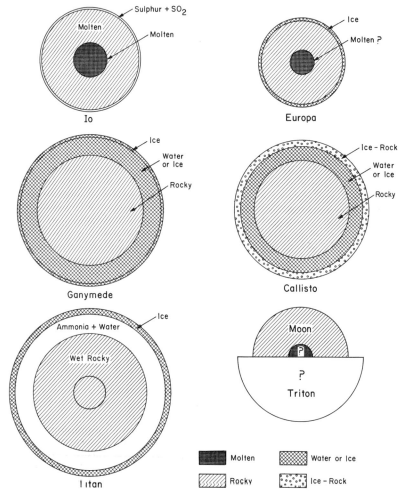

Fig. 191. Interiors of the major satellites. The core and rocky interior of Io are molten. The Galilean satellites are modeled after Torrence Johnson, Titan after John S. Lewis and Donald M. Hunten.

The story of Jupiter's Galilean satellites now constitutes a major subject of research. From Fig. 191 we see a progression of falling density with orbital radius. Clearly Ganymede and Callisto have basically our cometary composition, with some loss of the ices. Io and Europa, on the other hand, are very much Moon-like in gross structure. We have already dwelt on the heating of Io by tidal friction, induced by the other satellites and

Jupiter's equatorial bulge, whose gravitational disturbances force Io to move in an eccentric orbit about Jupiter. This tidal heating obscures the early history of Io because the icy materials would probably have been lost, even though Io may have originally been composed entirely of cometary ices and rocks. To some extent this may be true for Europa.

We are left, then, with at least three possible scenarios for the birth and development of the Galilean satellites. Two involve their growth in a cometary rotating nebula about Jupiter with two possible sequences. In one sequence they aggregated with more or less cometary composition but the inner ones were more heated by tidal friction and so lost their ices. In the other sequence, the temperature increased with decreasing distance from Jupiter, so that the ices failed to accumulate on Io but were more successful for Europa, Ganymede, and Callisto in sequence. In the third major scenario, the satellites were all captured in an extended Jupiter atmosphere or nebula.

We know so little about the smaller satellites of the major planets that discussion of their origin, capture, or *in situ* formation is speculative. Capture seems very likely for many or most of them.

Titan of Saturn and Triton of Neptune are not well enough observed to be treated thoroughly at present. They appear most likely to be of cometary origin but the retrograde motion of Triton so close to Neptune is unique. A capture scenario is strongly suggested in view of Nereid's direct orbital motion so far out from the planet. Is Pluto a lost satellite of Neptune and Triton a captured interloper?

Capture, incidentally, is rather difficult to explain for Deimos and Phobos of Mars. Celestial mechanics indicates that a small terrestrial planet would experience great difficulty in making such captures. Asteroidal collisions near the planet or formation in a ring system are the alternatives.

The Rings

The space-age discoveries of rings about Jupiter and Uranus have sharpened the wits of ring theorists. In 1848 E. Roche proved that fluid satellites, that is, bodies held together only by gravity, will break up when moving in orbits less than 2.44

$(\rho_p/\rho)^{1/3}$ times the radius of the planet, where ρ is the density of the body and ρ_p that of the planet. Although small fluid bodies are nonexistent, rocky bodies become weaker structurally with increasing size. Fracturing and crushing within asteroids, for example, make them more like fluids than rigid solids if subjected to tidal distortion. Note from Table 5 that the major rings of Saturn are within the Roche limit but that the satellite Pioneer 11 is barely outside. For a body with density of 1.3 times water, Roche's limit becomes 2.0 times Saturn's radius, leaving Pioneer 11 moderately safe from tidal disruption.

Generally, satellites cannot aggregate from small particles inside a distance comparable to the Roche limit. Nothing, however, prevents the collisional breakup of satellites at any planetary distance. Hence the ϵ ring of Uranus.

The great mass of the major rings of Saturn suggests to most astronomers that the rings are extremely old, perhaps, relics of unformed or broken satellites. Undoubtedly they are slowly losing some of their substance to Saturn by collisions and breakage coupled with various drag forces; at the same time they are growing from the interplanetary particles and gases that come their way. This may well be the origin of their icy coats. Their permanence and age, of course, are not directly observable nor derivable by present theory.

The multiple narrow dense rings of Uranus present a new phenomenon to the astronomical world. Theories are being developed, improved, and disproved too rapidly for a coherent explanation to be presented as yet. The most challenging problem is the eccentricity of the ϵ ring and the correlation of width with radial distance to Uranus. P. Goldreich and S. Tremaine conclude that the ring can maintain this configuration only by the self-containing gravity of the ring itself. They require a surface density of some 25 grams per square centimeter, adding up to a total mass of around 5×10^{18} gram. At a density twice water, the integrated mass would have a spherical diameter of 8.4 kilometers, not an excessive demand.

The celestial mechanics of ring formation and stability is far too complicated for exposition here. The Uranus rings extend to the Roche limit for component bodies of density 2.2 times water, consistent with the slow breakup of weak bodies and certainly unfavorable for aggregation. Thus each ring may represent the broken remnants of a tiny satellite, perhaps held in orbit by one or more sizable fragments. It is possible too that small unobserved satellites between the rings help to keep the ring particles close together by gravitational action.

In conclusion we note that the rings of Uranus may be somewhat transient in the sense that old ones may dissipate and new ones form. We have already noted the newly discovered ring of Jupiter is of an entirely different breed, the particles being microscopic and probably being replaced frequently. The ring may even be an extension of Jupiter's outer atmosphere.

Some astronomers now hold that *any* planet may develop rings, if only short-lived in astronomical time. John A. O'Keefe has even suggested that the Earth developed a prehistoric ring some 34 million years ago. The postulated ring shadowed the Earth and reduced the winter temperatures by some 20°C for 1 to 2 million years, staging the terminal event of the Eocene geological epoch. He places the source of the ring in a lunar volcano. Support for the speculation stems from a concurrent worldwide fall of *tektites,* small glassy pellets that melted as droplets somewhere, then cooled, and finally entered the Earth's atmosphere at low velocities. The collision of a small asteroid with the Earth, however, offers a more plausible explanation for this and other discontinuities in the geological record (see Chapter 6).

15

Origin and Evolution
of the Solar System

No longer can the philosopher in his easy chair expect to solve
the basic problems of the origin and evolution of the solar system. A great array of observational facts must be explained by a
satisfactory theory, and the theory must be consistent with the
principles of dynamics and modern physics. All of the hypotheses so far presented have failed, or remain unproved, when
physical theory is properly applied. The modern attack on the
problem is less direct than the old method, which depended
upon an all-embracing hypothesis. The new method is slower,
but it is much more certain. By a direct study of the facts, we can
specify, within an increasingly narrow range, the physical conditions under which the planets evolved. The manner of their origin must finally become apparent.

It is an interesting commentary on modern science that the
age of the Earth has been determined, although the details of its
origin remain a baffling problem. The oldest rocks in the
Earth's crust solidified nearly 4 aeons ago and the materials of

the Earth were brought together 4.6 aeons ago. Radioactive atoms within the rocks leave minute traces of lead, helium, and other atoms to constitute a measure of the time elapsed since the Earth cooled (Table 2, Chapter 6). Studies of meteorites and the lunar samples show that none have been solid for longer than the age of the Earth. Since meteorites are pieces of the solar system, we may conclude from their corresponding ages that the system is coeval with the Earth and Moon. The problem of the origin of the Earth is, therefore, synonymous with the problem of the origin of the system. Something happened 4.6 aeons ago to generate the planetary bodies and to produce the order and regularity that we observe today.

Unfortunately, the age of the Sun cannot be determined with the same accuracy as that of the Earth, Moon, and meteorites. The Sun may possibly be older than the planetary system. This leaves open the possibility that the planets were formed later, from interstellar material or from gases removed from the Sun or a passing star.

Outstanding orderliness is conspicuous in the planetary motions. The members of the solar system move in the same direction along a common plane. Not only do the planets and thousands of asteroids follow this plane in their revolution about the Sun, but the great majority of the satellites move about their primaries in a similar fashion. The Sun, moreover, and six of the nine planets exhibit the same phenomenon in their axial rotation. Even Saturn's rings share in the common motion. Of the few exceptions, we have mentioned the Uranus system, Venus, Neptune's Triton, and some of the outer satellites of Jupiter and Saturn. Pluto's axis of rotation is probably highly tilted.

The common motion of so many bodies suggests an initial rotary action, as though the solar system were once sent spinning by some cosmic finger. There is, in fact, so much motion in the outer bounds of the system that the older evolutionary hypotheses have failed in one respect: they cannot explain the *angular momentum* of the Sun compared with that of major planets. The angular momentum of a planet moving in a circular orbit at a given distance from the Sun (which is practically at the center of gravity of the solar system) is the product of its mass, distance, and speed. Since the speed diminishes only as the square root of the distance, a given mass contributes more angular momentum at a greater distance from the Sun. For a planet moving in an elliptical orbit, Kepler's law of areas (Chapter 2) expresses the constancy of the angular momentum at all times. When the planet is near the Sun it moves more rapidly than when it is farther away. No force toward or away from the Sun can change the angular momentum of a planet. Only an external push or drag along the orbit can increase or diminish this fundamental quantity of motion.

Jupiter, with its great mass, carries about 60 percent of the entire angular momentum of the solar system. The four giant planets together contribute about 99 percent, and the terrestrial planets 0.2 percent. The Sun, with a thousand times the mass of Jupiter, rotates so slowly that its angular momentum is only 0.5 percent of the whole. If all planets could be put into the Sun and could carry with them their present angular momentum, the augmented Sun would rotate in less than 10 hours, rather than a month.

A satisfactory theory for the origin of the solar system must first account for the existence of the planets, satellites, asteroids, and comets. It must explain how they were set moving in the remarkable manner already noted, and must provide the system theoretically with the observed amount of angular momentum. It must be flexible enough to explain the wealth of detailed knowledge we have gathered about all these bodies. Finally, developmental trends must be included to allow for the many changes that have taken place as these bodies have evolved.

In recent decades the measurement of the abundances of the elements and their *isotopes* (atoms of the same nuclear charges but of different masses, for example, hydrogen and heavy hydrogen or deuterium) gives us a powerful tool for establishing sources of materials. We can now eliminate from consideration the Sun or stars as sources. On Earth, in the atmosphere of Jupiter, and in interstellar gas the deuterium-to-hydrogen ratio is about 2×10^{-5}. Deuterium is rare but extremely stable unless subjected to great heat, as within stars, where it is quickly destroyed. The measured deuterium–hydrogen ratio in the Sun's atmosphere is only about 3×10^{-7}, or $\frac{1}{60}$ that for the planets and interstellar gas. We can immediately eliminate theories that make the planetary system out of gases ejected from the Sun, from passing stars, or from collisions of stars. The abundances of lithium can be used to support the same conclusion. Furthermore, as Lyman Spitzer showed, the quick scooping of gases from the Sun or a star would result in an explosive expansion. At the minimum scooping depth in a star the temperature of the gases is a million degrees or more, normally held there by the huge gravitational pressure. The sudden expansion by pressure release would prohibit all condensation or contraction processes required for planet or planetesimal formation.

A number of fascinating theories for planetary system evolution are now rendered obsolete, of historical interest only. They include first Georges L. L. deBuffon's 1745 theory that the Sun was sideswiped by a massive body (he suggested a comet). Modifications of this idea involving a passing star were proposed by Sir James Jeans (1877–1946) and Sir Harold Jeffreys to avoid the angular momentum problem of the Sun. In a final effort, Henry Norris Russell (1877–1957) proposed that a companion star of the Sun was struck and that the planets evolved from the debris. As we have seen, these theories are no longer acceptable. Nevertheless, one such theory by T. C. Chamberlin (1843–1928) and F. R. Moulton (1872–1952) involved an important concept, called the *planetesimal theory.* Here gases were released from the Sun by gigantic tides raised by a passing star and were sent spiraling about the Sun. The part that remained then condensed into small planetesimals, which finally agglomerated into the asteroids and planets. The concept of particle condensation and aggregation presented in the planetesimal theory is basic to several modern theories, even though the origin of the circumsolar cloud is entirely different.

Having eliminated the Sun and stars as sources of our planetary material, we are left only with the interstellar gas and dust. This source of material is consistent with our present knowledge of stellar formation. The question then remains as to whether the Sun and planetary system were coeval or whether the planets developed around the Sun out of captured interstellar material. If the Sun's age could be measured accurately, the question would be answered, but, alas, this is not the case. We have another clue, however, the measured ratio of abundant carbon of atomic weight 12 to rare carbon of weight 13. This ratio is about 90 in the Sun, on Earth, Venus, Mars, Jupiter, the meteorites, and in comets. Over the galaxy the ratio for carbon in the interstellar gas varies quite widely. The ratio is not much affected in the Sun by its rather mild thermonuclear reactions. If the measurements of the ^{12}C/^{13}C ratio were accurate to about 1 percent, we would have rather strong evidence that the entire solar system originated together. Unfortunately the accuracy is too poor for a definitive test.

Even though the rough constancy of the ^{12}C/^{13}C ratio is only indicative and not definitive, most investigators now favor the co-genetic origin of the Sun and solar system. Theoreticians find that it would have been very difficult for the isolated Sun to capture an adequate amount of interstellar gas and dust. The solar wind is effective in blowing away the extremely tenuous gas while the direct radiation similarly prevents the fine interstellar dust from collecting on or near the Sun. In moving through interstellar clouds, even relatively dense ones, the Sun by itself would maintain a clear volume near it. Consequently the Sun and its "hole" would simply plow through interstellar clouds like a jet aircraft through the stratosphere. Because no convincing theory for the accumulation of a planetary nebula about an isolated Sun has yet been presented, further discussion will center on theories involving the origin of the entire system from a collapsing interstellar cloud.

Interstellar clouds do not collapse easily because they normally contain widespread electric currents and magnetic fields. These fields, about $\frac{1}{100000}$ of the Earth's magnetic field, are evidenced by their ability to align the interstellar dust particles to polarize their scattered light. In other words, interstellar gas clouds are weak plasmas. The magnetic fields as well as the gas and light pressure tend to support them against collapse. Hence we observe stellar formation along the leading edges of spiral arms in our galaxy and others. Once instability has set in, C. Hayashi has shown that the collapse can occur rapidly, almost at the rate of free fall. The energy of collapse goes into dissociating the hydrogen molecules and then into ionizing them and helium, the two elements constituting almost all the mass. After the major collapse to a radius about the size of Mercury's present orbit, for some hundreds of thousands of years the star becomes several magnitudes brighter than its final magnitude will be when it settles down to a semistable system of energy generation such as operates in the Sun today. Remarkable discoveries of infrared stars by E. E. Becklin and by D. E. Kleinmann and F. J. Low demonstrate unquestionably the present-day occurrence of proto-solar systems in dust clouds.

G. Herbig has shown that the erratic T Tauri variable stars in dense gas–dust clouds are new stars, ejecting matter at a rapid rate. In fact, new stars more massive than the Sun eject an appreciable fraction of their mass in a stellar *gale,* as compared to our solar wind.

The observed new stars occur in regions where other stars are forming or have formed recently; exploding massive stars or *supernovae* leave traces in the abundance ratios of the elements. The study of meteorites, amazingly enough, leads to direct evidence that such was probably the case for the solar system. J. Reynolds found that the isotope xenon 129 occurs with unusually high abundance as compared to other xenon isotopes in certain meteorites. The only likely parent atom of xenon 129 is iodine 129, which is radioactive with a half-life of 17.2 million years. Since the xenon, a noble gas, must have been formed and held in the asteroidal body itself, the iodine 129 atoms must, therefore, have been created not too many tens of millions of years before the formation of the asteroids. The overabundance of the magnesium 26 isotope (normally 11 percent), found in fragments of certain meteorites, indicates the erstwhile presence of long-decayed atoms of aluminum 26, which has a half-life of only 0.7 millions years. The time allowed for the formation of the parent body, its breakup, and the aggregation of fragments into another body is reduced to only a few million years after the formation of the atoms. R. N. Clayton finds peculiar abundances of oxygen isotopes in certain meteorites, more evidence for supernovae activity in making atoms. Other evidence indicates that plutonium 244 (half-life of 76 million years) and perhaps super heavy radioactive elements were once present. All of this isotopic evidence supports the concept that our solar system was one of many formed in a huge stellar incubator. Perhaps a nearby supernova pressured our piece of the huge interstellar cloud, assisting in, or even causing, the collapse.

The electronic computer now is solving the problem of collapsing interstellar clouds, a problem that was intractable by older methods. R. B. Larson, a pioneer in this development, first showed how such a collapse can occur. We now know that huge clouds can fragment into smaller clouds, with stellar masses, and that double or multiple stars usually result. Irregular rings can form around a central concentration. But the computer has not yet solved the total problem, primarily because we still do not know enough to start the machine properly; we do not know physical details of interstellar clouds, particularly internal motions, magnetic field distributions, and so on. Nevertheless, these early calculations confirm our intuitive suspicions that the major problem is too much rotation or angular momentum, rather than too little, in the resulting primitive solar nebula after it contracts to dimensions of the present planetary system. Thus theorists who account for the solar system by assuming an original situation of extraordinarily low angular momentum are probably wrong. A better approach is to visualize processes that rid the system of excess angular momentum. Even so, the occurrence of sufficiently small rotation in the original cloud appears still to be rare.

In a sense we have almost returned to the hypothesis that has been believed for the longest time (excepting nonscientific speculations). It was presented apologetically by the great French mathematician Pierre Simon Laplace (1749–1827), at the end of the eighteenth century; it was somewhat similar to an idea of the noted philosopher Immanuel Kant (1724–1804). According to this *nebular hypothesis,* a rotating and therefore flattened nebula of diffuse material cooled slowly and contracted. In the plane of motion, successive rings of matter were supposed to have split off, to condense into the planets of our present solar system. Most of the matter finally contracted to form the Sun. Between the present orbits of Mars and Jupiter, the ring failed to "jell," and produced many asteroids instead of a planet. The sequence of events is pictured in Fig. 192. The Sun, according to the hypothesis, should have more angular momentum than the planets, not one-fiftieth as much.

The observational basis for the nebular hypothesis has since vanished. In Laplace's era the distances and natures of spiral nebulae and bright gaseous nebulae were unknown so that huge galaxies of stars (Fig. 193) and hot gaseous nebulae might legitimately have been considered examples of possible solar systems in early stages of evolution. We now look to dark ob-

Fig. 192. Laplace's nebular hypothesis. The condensation of a rotating gaseous nebula into the Sun, planets, and asteroids is here visualized. (Drawings by Scriven Bolten, F.R.A.S.)

scuring nebulae as potential regions for collapse into stars and possible solar-type systems. Even though Laplace's nebular hypothesis, as he presented it, is untenable, the present trend is to model it with modern physical concepts. One or more nebular rings may well develop in such a collapse but the multiring concept for forming the individual planets does not stand up theoretically. The major problem concerns the angular momentum.

One method for transferring angular momentum from the rapidly rotating Sun to the flattened solar nebula involves turbulence, or eddy currents. An attempt by C. von Weiszäker to accomplish this result theoretically is illustrated in Fig. 194. The concept is far too artificial to be accepted, particularly the required reverse rotation of the planetary condensations between the large eddies. The theory also is incapable of producing the slow, stately rotation of the Sun we see today. Nevertheless, turbulence theory can be evoked in a different fashion, as we shall see later.

Another process to slow the Sun's rotation involves the "wiry" magnetic field lines of highly ionized plasmas, the field of *magnetohydrodynamics* first initiated by H. Alfvén of Stockholm. The magnetic field lines certainly serve to prevent collapse of inter-

Fig. 193. The spiral nebula NGC 4736, photographed by the 200-inch reflector. Such a nebula contains more than a billion times the mass of the entire solar system, and therefore could not condense in the manner of Fig. 192. (Photograph by the Mount Wilson and Palomar Observatories.)

Fig. 194. Von Weiszäcker's eddies. Planets were presumed to form like ball bearings between the eddies and to rotate in the opposite direction. (From *Physics Today,* 1948.)

stellar gas clouds. Whether or not they persist throughout the collapse is controversial, but they can be expected to reappear in a rapidly rotating hot condensation at the center of the solar nebula.

Strong magnetic fields are present in the Sun, especially around sunspots. Radio noise from the Sun, and particularly the great outbursts of solar flares with associated cosmic rays and the solar wind, prove that in the Sun we are dealing with hot plasmas in which the embedded lines of magnetohydrodynamic force play a vital role. Young stars that will develop into the solar type show violent irregular variations of a similar nature. Since the late 1940s it has been clear that magnetohydrodynamics presents a likely process for transferring the angular momentum away from a rapidly rotating new star to leave a slowly rotating star like the Sun. In 1960 Fred Hoyle adopted a version of the concept in a theory of planetary evolution, starting when the entire proto-solar-system had collapsed to about the dimensions of Mercury's present orbit. He then connects the rapidly rotating Sun to the nebula by lines of magnetohydrodynamic force, analogous in some respects to long, elastic

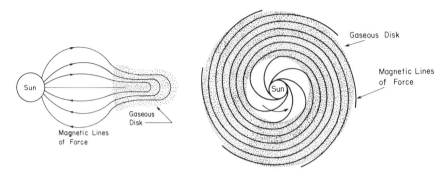

Fig. 195. Solar magnetic lines and the preplanetary disk, according to Hoyle. (*Left*) Solar magnetic lines of force connect across the plane of the ecliptic; the left-hand side of the "doughnut" is omitted. (*Right*) View normal to ecliptic shows winding of solar magnetic lines in the material of the preplanetary cloud.

threads tied to the ionized material within the planetary disk (Fig. 195). Since the outer disk is turning more slowly, the threads tend to wind around and around and stretch, thereby increasing the angular momentum of the disk and slowing down the rotation of the Sun. Unfortunately Hoyle's mechanism fails in transferring planetesimals to the outer regions of the planetary system. The tight magnetic spiral produces an *outward* pressure on the gases as well as a small forward component along their direction of motion. Consequently the gas moves as though it were under less than the normal solar gravitational attraction and hence more slowly than the planetesimals. The planetesimals, therefore, meet an effective resisting medium and spiral *inward* while the gases spiral *outward*.

It becomes clear generally that the sequence of solar-system evolution cannot involve an early highly condensed configuration with ensuing expansion to produce the planetary nebula out of which the planetary systems evolve. Such a scenario also involves an undesirably small amount of total angular momentum. We are left with some greatly modified version of the Kant–Laplace concept as the most acceptable.

In the previous chapter we found that the minimum amount of material of solar composition from which the terrestrial planets could have formed is some 200 to 250 times their present rocky mass. Hence an Earth mass was derived from about one Jupiter mass. The terrestrial planets and asteroids add up to perhaps 3 times this or 0.003 solar masses. The factor is about 50 for the icy-plus-rocky or cometary component. Each of the giant planets contains about 15 Earth masses, adding a minimum initial mass of perhaps 0.01 to 0.02 solar masses. The requirement for comets is quite uncertain, ranging from 1 to 100 Earth masses times 50, or 0.0002 to 0.02 solar masses. In all, the minimum initial mass required may be the order of 2 to 4 percent of the Sun's present mass, if we neglect the mass loss by the young Sun via the solar gale, which is perhaps another 5 to 10 percent. Efforts to produce the planetary system with the minimum initial mass lead to excessively long aggregation times and other difficulties, so that V. S. Safronov and B. J. Levin, for example, choose 1.2 solar masses as a more likely initial mass of the total system.

A. G. W. Cameron represents an extreme view in which the total rotating and contracting system once contained two or more solar masses, about half of which was lost to space. Other striking features distinguish his approach. Within the large mass of the extended flattened nebula, many protoplanets up to Jupiter's mass develop, having dimensions the order of an astronomical unit. They are initially large because Cameron grows the Sun and protoplanets simultaneously. At first the central mass is small and little concentrated at the center. As the interstellar material rains in, the mass and dimensions of the system grow to a maximum within 30 to 50 thousand years. The comets form in the outer regions at several hundred to one thousand astronomical units from the center.

Planet formation and loss of the solar nebula all occur very rapidly according to Cameron, in about 100,000 years after the cloud shrinks to about ten times the system's present dimensions. The transfer of nebular mass, outward from the growing inner solar region, takes place because of huge convective turbulence in the nebula. A beautiful theory for this turbulent mechanism has been developed by D. Lynden-Bell and J. E. Pringle. As mass is transferred rapidly outward by the turbulence, it carries away the excess angular momentum, some mass falling into the Sun. The Hayashi, or solar-gale, phase of the Sun finishes the task of removing the thin remaining nebula.

Because of the rapidity of the postulated mass loss the distant comets suddenly find themselves in a greatly reduced central gravity field and fly out into extremely elongated orbits with aphelion distances of tens of thousands of astronomical units. There they remain in deep freeze, the gravity of passing stars disturbing their orbital planes and occasionally sending one close to the Sun, as Jan Oort visualized in 1950.

Cameron's theory contains many new physical applications, too difficult to recount here. The protoplanets are mostly unstable, particularly against collision, because of their great size. This leads usually to the loss of most of their mass. In the case of the terrestrial planets this instability largely results from the nebula itself and the solar gravitational forces. The cores of the protoplanets that remain after the misadventures of the majority are hot as they emerge from the dissipating nebula. The small pieces cool quickly. Many collisions occur among these rocky bodies until the system becomes a somewhat orderly array of the surviving bodies and fragments. The terrestrial planets lose their primitive atmospheres and then gain their present atmospheres from internal degassing and infall of smaller bodies, including comets. The temperature in the nebula falls with solar distance, consistent with the evidence from meteorites, the planets, and the comets. Interstellar solids would mostly survive to within Venus' solar distance.

As Cameron's theory is continually developing with time, its current description will not be elaborated. Worrisome is the short time scale, more than an order of magnitude less than other theories. But, as we have noted before, the isotopic and radioactive records say little about time scales less than a few million years. The concept of large protoplanets, first suggested by Kuiper, must surely play an important role in all theories.

The processes of accretion and agglomeration in the solar nebula have been greatly clarified and the time scales reduced materially in a theory developed by P. Goldreich and W. R. Ward. It begins with a long-standing concept. In a nebula flattened by rotation, accretion of solids will begin with the growth of dust grains, perhaps minute interstellar grains, rocky or cometary depending on the temperature. As the grains grow, both by condensation from the gas and by sticky encounters, the grains will fall to the equatorial plane of the nebula. The gas will damp the motions of the grains to moderate both their collisional velocities and their tendency to overshoot the plane as

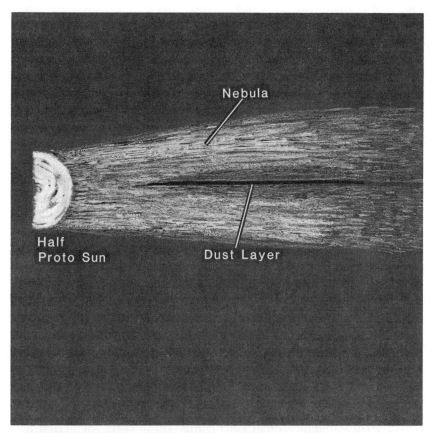

Fig. 196. Layer of small particles collected near the plane of the rotating flattened solar nebula. (Drawing by Joseph F. Singarella.)

they cross it in their orbital motions. Rather rapidly the grains will form a layer that is very thin compared to the thickness of the gaseous nebula, which is supported by the relatively high thermal velocities of the atoms and molecules (Fig. 196). Goldreich and Ward showed that gravitational instabilities within the plane of the thin layer of particles will draw them together much more rapidly than they would otherwise coalesce by slow encounters. Thus the time required to accumulate sizable bodies is greatly reduced.

We now have established the basis for what might be called the modern planetesimal theory. Although no theory is detailed enough to be accepted, and all may be basically wrong, the following sequence, in essence, is probably the most popular at the present time:

From the collapse of part of a great interstellar gas and dust cloud, very likely triggered by a nearby supernova, a flattened rotating solar nebula developed about a growing central Sun. Two, or perhaps several large protoplanets developed, finally to produce Jupiter and Saturn. In the flattened nebula, highly concentrated as compared with the interstellar cloud, the grains served as nuclei for condensation and grew. Within the solar distance of proto-Jupiter where the temperature rose, the grains may have lost their highly volatile ices, but well beyond proto-Saturn they were not heated seriously. Icy-rocky cometary bodies formed in these outer regions by accretions of grains—agglomeration—fell to the plane of the nebula, and finally collected in a thin layer, by gravitational attraction. Possibly smaller protoplanets of nebular composition served to collect these cometesimals into Uranus and Neptune. Pluto and Triton may indeed be effectively large comets. No estimates can be made as to the number of intermediate bodies that grew, then collided, or were swallowed up by larger ones.

The comets we see today may have been formed in this outer region and hurled into their huge orbits by perturbations, first by Uranus and Neptune, and some later by Saturn, Jupiter, and finally passing stars. Or, they may have been thrown out by Cameron's mechanism. Conceivably they were formed much farther out in smaller interstellar clouds gravitationally attached to the system.

Near the protosun, the temperature was high enough to melt most materials. Whether Mercury aggregated from iron silicate grains or is the remnant core of a sizable protoplanet remains for future research to decide. Between the Sun and Jupiter a huge number of planetesimals grew, aggregated, and collided in a maelstrom. The meteorites tell many stories of early times. All are fragments of bodies up to asteroidal size. Most of these fragments are made of lesser fragments of earlier bodies that were shattered by collisions. Strong magnetic fields were present. Tiny pieces have been bombarded by high-energy particles from the young Sun. Some meteorite sections show as many as 1000 million impacts per square millimeter. Some fragments are made of highly refractory materials, once certainly melted inside bodies at high temperatures. They are mixed often in a solid meteorite made primarily of low-temperature, relatively volatile materials, the carbonaceous chondrites. Many of the meteorites contain a large fraction of *chondrules,* glassy, round, or oval bodies up to a millimeter in diameter. These chondrules were once liquid drops that were rapidly cooled and then collected into large bodies. From the age records of the radioactive clocks, this frenzied activity occurred mostly in the first 200 million years, before the great bombardment of the Moon.

Some meteorites contain unusual volatile minerals that point to the presence of interstellar grains. It appears, therefore, that some, at least, of the primordial matter was not heated in the Jupiter-to-Sun region. This argues against all of the material having been processed in large protoplanets. On the other hand, G. J. Wasserburg and his colleagues at the California Institute of Technology have demonstrated an excess of the magnesium 26 isotope in certain tiny fragments or *inclusions* in a meteorite. The only reasonable source is aluminum 26, which has a half-life of only 740,000 years. The aluminum must have been made in a supernova and the atoms incorporated into sizable asteroid bodies, all within a short time, less than about three million years. Then the bodies had to cool partially, be broken by collisions, and the pieces had to again accumulate with others into the parent asteroidal body of the meteorite. Early heating of such bodies is easily explained because aluminum 26 was abundant enough to have provided sufficient energy if the required time was short. Other short-lived supernova elements could have helped. On a longer time scale potassium 40 (half-life 1.3 aeons) is adequate. Also C. P. Sonnett has suggested that the solar gale may have heated protoplanets to depths of the order of 100 kilometers. The reality of the supernova now appears to be a certainty. Its role in initiating or accelerating solar-system formation still remains conjectural. The meteorites suggest a maximum of about ten million years for much of the asteroidal formation, but the terrestrial and lunar formation might have required up to a hundred million years, the 0.1 aeon uncertainty mentioned earlier.

The number, sizes, and collisional history of the planetesimals is lost information, never to be recovered in detail. Of particular interest is the regularity of orbit spacing in the inner solar system, represented by Bode's law (Appendix 1). The electronic computer now enters the field again. Computer runs, starting with a number of variously sized planets and allowing for collisions, gravitational effects, and ejections, show that the present distribution of planets and orbits is not an unreasonable outcome from quite different beginnings. In any case, the older picture of each planet growing in its present orbital confines is not necessarily correct. Planets, like animals, appear to require a certain amount of living space. Indeed, a certain analogy with biological evolution is appropriate. In both planetary and biological evolution we primarily observe and study the survivors! Also we observe a fossil record of many lost biological species and a few remnants of planetesimals. Natural selection in the planetary case favored the bodies that, by chance, grew most rapidly at the beginning. They tended to accelerate their growth by the assimilation of smaller bodies.

The solar nebula, whatever its mass, was surely dispersed fairly soon, on the order of a million years after the major accumulation. Presumably both turbulence and magnetohydrodynamic effects were responsible. The young Sun brightened strikingly and perhaps irregularly and blew off an appreciable percentage of its mass in a solar gale lasting a few million years. Finally it settled down to a steady radiance of possibly half its present value. In this period the Sun lost most of its angular momentum and slowed its rotation by the plasma braking effect of the solar gale. This braking still continues but at an extremely low rate.

The prevalent prograde rotations of the planets is not a simple obvious consequence of rotation in the solar nebula. The asteroids are spinning at random, probably a consequence of many random collisions. The planets and associated satellite systems are mostly spinning prograde. Mercury has probably lost its rapid rotation by tidal friction with the Sun, while Venus may well have been rotating rapidly retrograde along with a long-lost satellite. Tidal friction could have removed the satellite and slowed the rotation. For Uranus, a late hit by a body possessing some 10 percent of the planet's mass could have tilted its rotation axis. If it then possessed a ring of moonlets, their plane would have tilted to the plane of the new equator by the perturbations of its equatorial bulge coupled with collisions among the moonlets. Similar events could have occurred for other planets with developing satellite systems.

Protoplanets with a very large sphere of potential capture should attain prograde rotation. Among small accreting planets the eventual sense of rotation is very sensitive to the orbital eccentricities of the planetesimals and their distribution in solar distance. Encounters with relatively large planetesimals can produce large effects on their axes of rotation. After a small planet has largely cleaned out a ring or band near its orbit, it will probably be rotating fairly slowly in an unpredictable sense. Further accretion derives from planetesimals in eccentric orbits mostly near the edge of this band. Those in outer orbits will strike the planet near their perihelia with higher velocities than the planet. Conversely those from inner orbits will strike near their aphelia with lower velocities than the planet. If the number of planetesimals near the edge of the band increases with distance from the planet's orbit, both within and without, more of the outer planetesimals will strike the hemisphere of the planet away from the Sun and hence add prograde momentum. Conversely, the sunward hemisphere will tend to overtake more of the inner planetesimals, again to increase the prograde rotation. A relatively large planetesimal can, of course, upset the final result. All the planetary equators are, indeed, tilted appreciably to their orbit planes and also to the mean plane of the solar system.

Why little asteroids rather than a sizable planet were formed between Mars and Jupiter remains unexplained. Probably Jupiter is responsible. Forming early in the nebula, the great mass of Jupiter may have disturbed the motions of the nearby gas, dust, and planetesimals so that accretion processes were slowed down. Possibly a protoplanet in the region was perturbed to fall into Jupiter, or kicked out of the system, or torn apart in the making. Once the solar gale began to dissipate the hydrogen and helium of the solar nebula, however, the many asteroids began to disintegrate slowly by collisions, as they do today. We have no way to estimate the total number or mass, which could have been enormous or small.

Four possible mechanisms for producing the Moon are seriously discussed today. The most obvious is that many "moonlets" collected in a ring near the Earth. The tidal friction caused a larger and inner moonlet to spiral outward, growing as it spiraled and swallowing up all of the outer ones. The remaining inner ones, being small, were dissipated away, perhaps by the violence that removed the Earth's original atmosphere, perhaps by perturbations, or perhaps by collisions with debris. This "obvious" method of making the Moon gives no clue as to its very low mean density; the composition of the Moon argues against the theory.

A second possibility is that the Moon was captured, readymade. A mechanism of capture, developed by H. Gerstenkorn, is not impossible. But why should the Moon's density be lower than that of the other bodies formed in this region of the solar system?

A third possibility is a modern version of Sir George Darwin's tidal separation of the Moon from the Earth. Suppose, with D. U. Wise, that when the Earth formed it was rotating very rapidly, near the limit of stability. As radioactivity melted the interior, the dense iron settled to form the core. The total angular momentum, of course, remained constant but the mantle became less dense and contained a smaller fraction of the total mass. As a consequence the Earth turned faster. Thus the atmosphere and the outer mantle near the equator were thrown out to maintain stability. Probably in such circumstances, the Earth would have elongated something like a tenpin and suffered fission by throwing out the Moon mass from the small end. Tidal friction then moved the Moon to its present position. This fission theory of the Moon is highly attractive because it forms the Moon out of the lower density mantle of the Earth and simultaneously removes the Earth's primitive atmosphere. The theory, however, lies on the verge of the impossible because of the limited effects arising from the settling of the iron core.

Perhaps the theory currently most popular is a somewhat grazing collision of a Moon-sized planetesimal with the Earth. Material from both the Earth and the invader would accumulate in a ring around the Earth, repeating the above scenario of the "moonlet" theory. Much of the material would be greatly heated in the process, accounting for the lack of volatiles in Moon rocks. Much of its composition would be Earth mantle, consistent also with the Apollo results. Probably computer simulations will finally clarify the probability of this solution for the Moon's origin.

The detailed formation of the other satellites in the solar system remains a subject fraught with great uncertainty. The current trend is to believe that satellites near the planets, revolving in the same sense as the planets, are indeed miniature solar systems, the satellites having grown in flattened "nebulae" around their planets. The low densities of the outer Galilean satellites suggest that an overflow of "comets" from the Uranus–Neptune region were caught in the outer fringes of proto-Jupiter or that the temperature dropped low enough for awhile to produce ices as close to the Sun as Jupiter. We have already noted that the inner ones, Io and Europa may either have been formed mostly from rocky material or may have lost their volatiles by tidal heating.

The origin of Neptune's Triton remains a mystery; it moves near the planet in a retrograde orbit. Was it an interloper that displaced the original satellite, now Pluto? Could it possibly have been captured by the enlarged proto-Neptune, or by collision with another large satellite? Could a collision have produced a retrograde ring that aggregated into Triton? Space missions to Neptune will probably tell the story.

The atmospheres of the giant planets present no serious evolutionary problem, as we have accounted for the appreciable hydrogen–helium contents of the planets themselves. These light gases will simply rise to the surface, while the "icy" content accounts for the clouds, methane, ammonia, and other molecules. The heat generation within Jupiter, Saturn, and Neptune is a valuable guide to their interiors. Evolution is not yet completed for these planets, nor probably for Uranus. The slow separation of the heavier atoms, molecules, and minerals toward their centers seems an adequate explanation for the energy generation and is a help in understanding the physics at great pressures and high temperatures.

We have already discussed the atmospheres of the Earth, Venus, and Mars. Mars has clearly lost a great deal of its secondary atmosphere. The solar gale, aided perhaps by a very early hot surface, blew away the primitive nebular atmosphere. The nitrogen 15 to nitrogen 14 ratio is about 60 percent greater than the terrestrial value, suggesting a great loss of the secondary atmosphere from the outer atmosphere. However, the xenon 129 to xenon 132 ratio and the argon 40 to argon 36 ratio are much larger than for the Earth. M. Shimizu attributes these differences to pollution of Mars by supernova products carried to Mars by meteoritic bodies. This story is not yet complete.

The Venus atmosphere contains a huge quantity of carbon dioxide but little water. As we have seen, the Earth's atmosphere once held a comparable amount of carbon dioxide, the carbon having been fossilized in carbonate rocks. But did Venus once have as much water as the Earth? Only rather artificial arguments have been put forward to account for the lack of water in the secondary atmosphere of Venus. Perhaps the higher temperature closer to the Sun caused the interior of Venus to be degassed more rapidly than Earth. But then why so much carbon dioxide? More popular is the runaway greenhouse theory in which Venus once held as much water as the Earth. Again, because Venus is closer to the Sun, the greenhouse effect from carbon dioxide heated the lower atmosphere much more in early times than today. The water was rapidly photoionized in the high atmosphere, the hydrogen lost to space. The oxygen was then combined in the rocks of Venus. The oxygen content of the Earth's atmosphere is generally attributed to life forms, the atmosphere once being essentially free from oxygen; primitive life forms appear to have been anaerobic. (Figure 197, after Carl Sagan, portrays possible surface temperatures on Earth, past and future.)

Mercury, Mars, the Moon, and many of the satellites are dead. They have ceased to evolve. They will simply cool very slowly; essential changes will come only by accidental collisions and forced orbital changes. After many aeons the Sun will brighten and then fade, leaving the planets in the cold. Tidal friction will probably carry the Moon outward until the month and day are equal, one side of the Earth continuously facing the Moon. This common period will be much longer than the present month. Then the Moon will slowly spiral inward. The Earth and Venus are still evolving internally, but the radioactivity is slowly dying away and they are losing heat to space. Over many aeons they too will become cold spheres, with frozen atmospheres, or else suffer accidents.

Many problems of solar system evolution remain unmentioned or untreated in this brief review of the subject, but research progress is extremely rapid. Direct space exploration

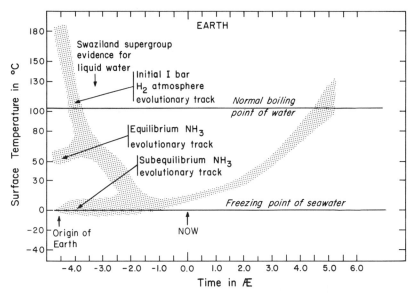

Fig. 197. A plot of surface temperature versus time in billions of years, predicting the future of the solar system. (Courtesy of C. Sagan.)

coupled with the unbelievable sensitivity of chemical and isotopic analysis is providing tools of such power that clear-cut answers to many old problems are in sight. Furthermore, the observations of infant stars in stellar nurseries give us both confidence and insight concerning hitherto baffling aspects of planetary evolution.

The question as to the numbers of planets about other stars remains unresolved. Current opinion strongly favors the concept that almost all stars form in essentially the same fashion. Only in one respect is the Sun atypical; it is a single star having planets. A majority of stars appear to be double or multiple. Most double stars cannot hold stable systems of nearby planets, although some could do so. It is not at all unlikely that one star in a hundred may be single and thus *able* to possess planets. In our own galaxy of 10^{11} stars, it is not unlikely that one out of a million to perhaps one out of a thousand stars may have a planet like the Earth moving under conditions that are comparable to the solar radiation on the Earth. This leaves us in our galaxy perhaps 1 to 1000 million planets on which sentient beings *might* develop; and there are millions of galaxies.

Few doubt that life forms develop where circumstances are favorable. The time of development for sentient beings, however, is extremely long, at least as evidenced in our fossil records. Man has been on this planet for less than 0.1 percent of its existence. If he can persist for 100 million years, 2 percent of the present age of the Earth, many skeptics will have erred. If so, however, there is a chance that perhaps 10,000 to 10,000,-000 planets in our own galaxy contain sentient beings more or less like ourselves. Space travel to visit them is now virtually impossible in view of the limitation of travel imposed not only by Einstein's theory of relativity but by laboratory experience. We and our machines are made solely of atomic nuclei and electrons. In high-energy accelerators both types of particles are given enormous energies that would be sufficient under Newtonian theory to set them moving at huge multiples of the velocity of light. The multiple is about 100 times for protons and electrons, millions in cosmic rays. The atomic nuclei and electrons *cannot* be made to travel faster than light. As their energy increases they approach this critical velocity while their mass increases precisely according to relativity theory. Our high-energy accelerators would fail to operate if this were not true. Since we cannot make our component parts move faster than light, it is obviously impossible for us to make our bodies do so, or our machines.

As for the same limitation on signals, however, a possibility has been suggested by G. Feinberg. He theorizes that there may be particles, *tachyons,* that cannot move *slower* than light. His postulated tachyons lose energy as they speed up. Thus relativity theory may conceivably allow signals to be sent at hyper-light velocities, even though ordinary matter will always be chained. If tachyons can first be demonstrated and then controlled, they present a possibility for communication with other intelligent beings in the universe. The other possibility for such communication within the foreseeable future is the remote chance that such beings might wish to signal us by radio or the greater chance that they might use such powerful transmitters in their own communications that our great radio telescopes could intercept their signals. A serious attempt to listen for such signals has been conducted by the National Radio Astronomy Observatory at Green Bank, West Virginia, under the code name

Project OZMA. A great CYCLOPS project involving square kilometers of radio antennas has been proposed. The arguments for and against the allocation of great effort in such programs are thought provoking. Much as I like the imaginative, science-fiction aspects of such programs I must express some skepticism regarding their success in the near future. Still, there *must* be other intelligent beings in the universe. They, however, may have developed less energy-consuming modes of communication than our primitive science of today.

As for interstellar travel, the limitations both on energy and on velocity make a manned round trip to the nearest star, 4.3 light years away (see Appendix 4), an operation that would span generations. Besides, with today's technology, it might create a budgetary catastrophe. Such a venture is not impossible but it is fantastically improbable until more powerful and much cheaper power sources become available. Fusion power, however, may be just beyond today's horizon. Success in detecting extraterrestrial cultures or, better, in communicating with them would, of course, crystalize world interest and might trigger a colossal international effort to explore interstellar space. Nothing, at least in my view, could be more exciting or more stimulating to the human race.

We have studied the present state of our planetary system and, to a limited extent, its history. Its future, unless some unforeseen accident occurs, seems bright. The chance that a wandering star might disrupt the stately order of the planetary motions is small, even within aeons. Nor should we expect a great change in the Sun's radiance much sooner. Probably the glacial ages will recur; we cannot say. Man may change the climate, but having done so, he may reverse his mistakes. The continents may rise and fall during the ensuing ages, as they have done in the past—we hope they do it slowly. And random meteoric masses will pierce the surface here and there.

But order, which the solar system has finally attained, will prevail.

Appendix 1

Bode's Law

The so-called Bode's law, ascribed to J. E. Bode (1747–1826), is not a physical law but only a convenient rule for recalling the distances of the planets from the Sun. Write down a series of 4s, one for each planet. Add to the successive 4s the numbers 0 for Mercury, 3 for Venus, 6 for the Earth, 12 for Mars, 24 for the asteroids, and so on. Insert a decimal point in each sum to divide by ten. The resultant series of numbers represents approximately the distances of the planets from the Sun, in astronomical units. The scheme of numbers follows:

	Mer.	Ven.	E.	Mars	Ast.	Jup.	Sat.	Ur.	Nep.	Pl.
	4	4	4	4	4	4	4	4	—	4
	0	3	6	12	24	48	96	192	—	384
Bode's law	0.4	0.7	1.0	1.6	2.8	5.2	10.0	19.6	—	38.8
Actual	0.39	0.72	1.00	1.52	—	5.20	9.54	19.18	30.07	39.67

Note that Bode's law includes the asteroids and gives the distance for Pluto rather than Neptune. The law was used in Leverrier's and Adams' predictions of the position of Neptune. The predicted orbits were therefore considerably in error.

No theoretical basis for the rule has been generally accepted.

Appendix 2

Planetary Configurations

The various geometric positions of the planets with respect to the Sun and the Earth are known as the *planetary configurations*. They are shown in Fig. 198. For an observer on the Earth the angle between a

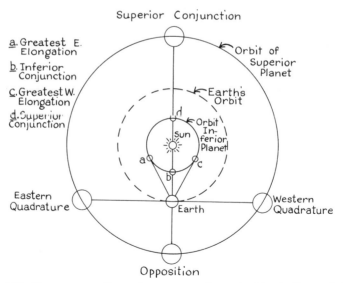

Fig. 198. Planetary configurations as seen from the Earth, for an inferior and a superior planet.

planet and the Sun is the *elongation*. A *superior* planet, whose orbit lies outside the Earth's, passes through all elongations to 180° east or west. An *inferior* planet, whose orbit lies within the Earth's, may attain only a certain maximum elongation: the *greatest eastern elongation* or *greatest western elongation*. An inferior planet comes to *inferior conjunction* when passing the line between the Earth and the Sun, and to *superior conjunction* when in line with the Sun beyond it.

A superior planet also may come to *superior conjunction* (or simply to *conjunction*). When directly opposite to the Sun, the configuration is *opposition*. The configurations at right angles to the Sun's direction are *eastern* or *western quadrature*.

An inferior planet is best observed near greatest elongation, east or west. A superior planet is best observed at opposition.

Appendix 3

Planetary, Satellite, and Other Data

The following comments on the data of Table 6 are numbered to correspond to the number of the column.

1. One astronomical unit is the mean distance from the Earth to Sun. It is 149,597,900 kilometers (92,955,800 miles). The mean distance from the Earth to the Moon is 384,401 kilometers (238,856 miles); maximum 406,700 kilometers; minimum 356,400 kilometers.

2. The sidereal period is the time of one revolution with respect to the stars. One tropical year (of the seasons) is 365 days 5 hours 48 minutes 46.0 seconds. It is the unit of columns 2 and 3.

3. The synodic period is the time of one revolution with respect to the Sun as seen from the Earth.

4. For definition see page 26.

6 and 7. Mean values are given. The Earth's equatorial diameter is 12,756.3 kilometers (7926.4 miles) and its polar diameter is 12,713.6 kilometers (7899.8 miles).

10. The Earth weighs 5,976,000,000,000,000,000,000 metric tons (5.976×10^{21} tons).

12. The weight of a given object on the Earth when multiplied by the quantity in Table 6 becomes the weight of the object at the surface of the planet.

TABLE 6. *Planetary data.*

Body	(1) Mean distance to Sun	(2) Sidereal period	(3) Synodic period	(4) Eccentricity of orbit	(5) Inclination of orbit to ecliptic	(6) Orbital velocity (km/sec)	(7) Equatorial diameter (km)	(8) Polar flattening	(9) Volume (Earth = 1.0)
Mercury	0.387	87^d97	115^d88	0.206	$7^\circ0$	47.9	4,878	?	0.056
Venus	0.723	224^d70	583^d92	0.007	$3^\circ4$	35.0	12,100	?	0.858
Earth	1.000	365^d256	—	0.017	$0^\circ0$	29.8	12,756	1/298.35	1.000
Moon	1.000	27^d32	29^d53	0.05	$5^\circ1$	1.03	3,477	1/2000	0.0203
Mars	1.524	687^d0	779^d9	0.093	$1^\circ8$	24.1	6,790	1/120?	0.150
Jupiter	5.203	11^y86	1^y092	0.048	$1^\circ3$	13.1	142,700	1/15.4	1,313
Saturn	9.539	29^y46	1^y035	0.056	$2^\circ5$	9.6	120,000	1/11	759
Uranus	19.18	84^y01	1^y012	0.047	$0^\circ8$	6.8	51,800	1/40?	66
Neptune	30.06	164^y8	1^y006	0.009	$1^\circ8$	5.4	49,000	1/50?	58
Pluto	39.44	247^y7	1^y004	0.250	$17^\circ2$	4.7	3,000	?	0.01?
Sun	0	—	—	—	—	—	1,391,000	—	1,300,000

TABLE 6 (*continued*)

Body	(10) Mass (Earth = 1.0)	(11) Density (water = 1.0)	(12) Surface gravity (Earth = 1.0)	(13) Velocity of escape from equator (km/sec)	(14) Period of rotation	(15) Maximum surface temperature (°C)	(16) Number of satellites	(17) Albedo	(18) Inclination of equator
Mercury	0.055	5.44	0.38	4.25	58.65	430	0	0.056	$<30^\circ$
Venus	0.815	5.25	0.906	10.37	243^d	465	0	0.72	177°
Earth	1.000	5.52	1.0	11.20	23^h56^m	60	1	0.39	$23^\circ5$
Moon	0.01229	3.34	0.165	2.37	27^d3	101	0	0.07	—
Mars	0.1075	3.95	0.38	5.0	24^h6	20?	2	0.16	$24^\circ0$
Jupiter	317.9	1.34	2.6	60	9^h8	−144	14	0.70	$3^\circ1$
Saturn	95.1	0.70	1.07	35	10^h2	−174	11	0.75	$26^\circ7$
Uranus	14.6	1.23	0.88	21	16?	−215	5	0.90	$97^\circ9$
Neptune	17.2	1.64	1.14	24	16?	−218	2	0.82	$28^\circ8$
Pluto	0.003?	~1?	0.05?	~1?	6^h4	?	1?	0.25?	?
Sun	332,800	1.41	28	618	25^d	5,400	?	—	$7^\circ2$

TABLE 7. Satellite Data.

Satellite	(1) Diameter (km)	(2) Mass (Moon = 1)	(3) Density (H$_2$O = 1)	(4) Planetary distance (1,000 km = 1)	(5) Period (days)	(6) i	(7) Discovery
Earth							
Moon	3,476	1.000	3.341	384	27.322	23°	Unknown
Mars							
Phobos	22*		~3	9	0.319	1	Hall, 1877
Deimos	13*		~3	23	1.262	2	Hall, 1877
Jupiter							
XIV —	20:			102?	0.210?	0?	Voyager, 1979
V Amalthea	346*			181	0.498	0	Barnard, 1892
I Io	3,640	1.214	3.53	422	1.769	0	Galileo, 1610
II Europa	3,130	0.663	3.04	671	3.551	1	Galileo, 1610
III Ganymede	5,280	2.027	1.93	1,070	7.155	0	Galileo, 1610
IV Callisto	4,840	1.447	1.79	1,880	16.689	0	Galileo, 1610
XIII Leda				11,100	239	27	
VI Himalia	170:			11,500	250.6	28	Perrine, 1904
VII Elara	80:			11,700	260	25	Perrine, 1905
X Lysithea	26:			11,900	264	29	Nicholson, 1938
XII Anake	22:			21,700	631	147	Nicholson, 1951
XI Carme	28:			22,500	692	164	Nicholson, 1938
VIII Pasiphae	24:			23,500	739	145	Melotte, 1908
IX Sinope	26:			23,700	758	153	Nicholson, 1914

Saturn							
XI	100:			152	0.70	0?	Pioneer, 1979
X Janus??	220:			169	0.815	0	Dollfus, 1966
I Mimas	400:	0.00051	~1?	186	0.942	2	Herschel, 1789
II Enceladus	500:	0.00115	~1?	238	1.370	0	Herschel, 1789
III Tethys	1,000:	0.0085	~1?	295	1.888	1	Cassini, 1684
IV Dione	1,150:	0.0158	~1?	378	2.737	0	Cassini, 1684
V Rhea	1,600:	0.0025?	~1?	528	4.518	0	Cassini, 1672
VI Titan	5,840:	1.905	1.34	1,223	15.95	0	Huyghens, 1655
VII Hyperian	440:	0.0014?		1,484	21.28	0	W. Bond, 1848
VIII Iapetus	1,500:	0.0305?	~1?	3,563	79.33	15	Cassini, 1671
IX Phoebe	200:			12,930	550.4	150	Pickering, 1898
Uranus							
V Miranda	400:	0.0014?		130	1.414	3	Kuiper, 1948
I Ariel	600:	0.0177?		191	2.520	0	Lassell, 1851
II Umbriel	250:	0.0068?		266	4.144	0	Lassell, 1851
III Titania	1,000:	0.0585?		436	8.706	0	Herschel, 1787
IV Oberon	900:	0.0354?		583	13.46	0	Herschel, 1787
Neptune							
I Triton	4,480:	1.9	2?	356	5.877	160	Lassell, 1846
II Nereid	240:			5,570	359.9	28	Kuiper, 1949
Pluto							
Charon?	1400?	0.02?		20?	0.266?	?	Christy, 1978

13. An object shot away from the equator with this velocity would escape forever into space (neglecting friction with an atmosphere).

14. The Sun's period of rotation varies from 25.0 days at its equator to 26.6 days at latitude 35°.

18. The *albedo* is the ratio of the total amount of light reflected by the planet to the light incident on it.

19. The inclination of the equator is given with respect to the orbit plane of the planet (Moon and Sun to ecliptic).

The following comments on Table 7 are numbered to correspond to the number of the column.

1. A colon indicates great uncertainty as the diameter is estimated only from the magnitude. An asterisk indicates that the diameter given is the cube root of the product of the three axes.

2. Unit is the Moon's mass, 7.35×10^{25} grams or 7.35×10^{19} tons.

3. Unit is water or 1 gram per cubic centimeter.

4. Measured from the center of the planet in units of 1000 kilometers.

5. In mean solar days or 24 hours.

6. Inclination, i, to plane of planet's equator. If i is greater than 90°, the motion is retrograde.

7. Discoverer and date of discovery.

Miscellaneous Data

Velocity of light: 299,792.5 km/sec = 186,282.4 mi/sec

Constant of gravity in Newton's law, Force = Gm_1m_2/r^2: $G = 6.670 \times 10^{-8}$ cm³/(gm sec²)

Parallax of the Sun: 8.7942 arc sec

Time required for light to travel 1 astronomical unit: 499.005 sec

Light year the distance light travels in one year: 9.4605×10^{17} cm = 5,878,500,000,000 miles

Surface gravity of the Earth at the equator: 978.031 cm/sec² (3.392 cm/sec² centrifugal acceleration)

Rotational velocity of Earth's equator: 0.4651 km/sec. = 1040 mi/hr

1 meter = 39.37 inches precisely = 1.0936 yards
1 kilometer (km) = 0.621371 mile; 1 mile = 1.60934 km

1 kilogram (kg) = 1000 grams = 2.20462 pounds (lb)
1 metric ton = 1000 kg = 2204.62 lb = 1.10232 tons (ordinary)

Temperature in degrees Centigrade (°C) = $\frac{5}{9}$ (Temp. °F − 32°)
Temperature in degrees Fahrenheit (°F) = $\frac{9}{5}$ (Temp. °C) + 32°
1 bar = Atmospheric pressure at the surface of the Earth
1 meter per sec. = 2.237 miles per hour

Appendix 4

The Star Chart

The accompanying star chart (Fig. 199) is intended primarily for use with Table 9 (Appendix 5) to locate and identify the planets at any time from 1980 through 1990. A chart of the north polar region is shown in Fig. 200.

The *magnitudes* of stars define their brightness on a reversed scale. A first-magnitude star has the average brightness of the 20 brightest stars in the sky. A sixth-magnitude star is just one-hundredth as bright, and can barely be seen with the naked eye on a very clear dark night. Each magnitude denotes a step of 2.512 times ($\sqrt[5]{100}$) in brightness. Thus a star of the sixth magnitude is 2.512 times as bright as one of the seventh magnitude, and a hundred times as bright as one of the eleventh magnitude.

For the most brilliant stars the values become negative. Sirius, the brightest star in the entire sky, is of magnitude −1.43. Venus, the brightest planet, sometimes reaches a magnitude of −4.3. It is then more than a hundred times brighter than a first-magnitude star. Jupiter, at maximum, reaches a magnitude of −2.5, Mars −2.8, Saturn −0.4, and Mercury −1.2. Uranus is of magnitude 5.7, theoretically visible to the naked eye, but seen by very few people. Neptune is of

THE STAR CHART

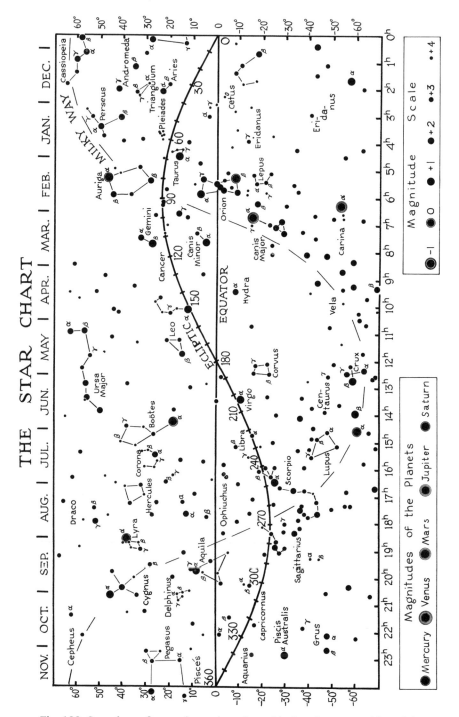

Fig. 199. Star chart, for use in conjunction with the planetary tables of Appendix 5.

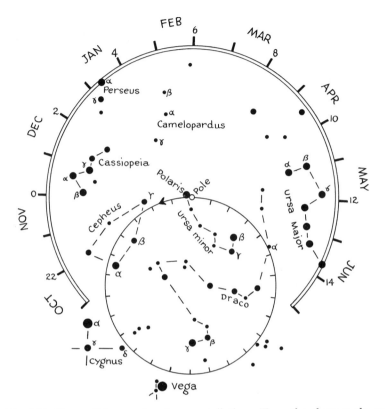

Fig. 200. Chart of the north polar constellations. Turn the chart so the correct month is uppermost and face north for observations near 8 p.m. The smaller circle shows the path of the pole of the ecliptic due to precession of the equinoxes. Each mark on the circle represents 1000 years and the whole motion is completed in about 26,000 years. In A.D. 14,000 Vega will be the pole star. (Chart by Donald A. MacRae.)

magnitude 7.6, easily visible in quite a small telescope, but more difficult to detect than Uranus.

Venus, Jupiter, and Mars are brighter than any star, hence easy to identify in the sky. Saturn is also easily visible, as only a few stars exceed it in brightness. The planets may sometimes be identified by the steadiness of their light; the stars twinkle much more violently. Mercury is always so near to the Sun that no attempt should be made to find it except at its greatest elongation, and then only under favorable observing conditions.

The constellation identification for groups of stars arose in ancient times; the names are mostly Latin, Greek, and Arabic. In recent years the boundaries of the various constellations have been assigned by international agreement. The brightest star in a constellation is gen-

TABLE 8. *The brightest stars.*

Name		Visual magnitude	Distance (light years)
Constellation	Common		
α Canis Majoris	Sirius	−1ᵐ43	9
α Carinae	Canopus	−0.73	500?
α Centauriª	Rigil Kent	−0.27	4.3
α Bootis	Arcturus	−0.06	32
α Lyrae	Vega	0.04	26
α Aurigae	Capella	0.09	45
β Orionis	Rigel	0.15	700?
α Canis Minoris	Procyon	0.37	11
α Eridani	Achernar	0.53	140
β Centauri	Agena	0.66	140
α Orionis	Betelgeuse	(0.7)	270
α Aquilae	Altair	0.80	16
α Tauri	Aldebaran	0.85	68
α Crucis	Acrux	0.87	160
α Scorpii	Antares	(0.98)	400
α Virginis	Spica	(1.00)	230
β Geminorum	Pollux	1.16	34
α Piscis Australis	Fomalhaut	1.16	23
α Cygni	Deneb	1.26	1600?
β Crucis	Becrux	1.31	470?

a. Its fainter companion, Proxima Centauri, is the nearest star.

erally called alpha (α), the second beta (β), the third gamma (γ), and so on through the Greek alphabet, although there are many exceptions. (The possessive case of the constellation name is used in such designations.) Many of the stars have proper names as well. The most conspicuous constellations are labeled on the chart and the stars α, β, and γ are designated in many of them. The chart is complete to the third magnitude (with only two or three exceptions) and shows a number of the fourth-magnitude stars.

The names of some of the brightest stars are given in Table 8.

The horizontal line across the middle of the chart is the *celestial equator* on the sky. In the United States it can be located when one faces south and looks up at an angle of from 40° to 65° (90° minus the observer's latitude).

The long curved line that crosses the celestial equator is the *celestial ecliptic*. The planets all appear within about 7° of the ecliptic. The broad band, with the ecliptic as its central line, that includes the paths of the planets is called the *zodiac*. The numbers along the ecliptic give, in degrees, the *celestial longitude*, corresponding on the sky to ordinary longitude on the Earth. (Table 9 gives the celestial longitudes of the planets at convenient intervals of time from 1981 through 1992.) The scale along the bottom of the chart is the *right ascension,* measured in hours of time; 24 hours =360°, therefore 1 hour =15°. Right ascension is like longitude except that it is measured along the celestial equator instead of along the ecliptic.

The vertical scale at the sides of the chart gives the *declination.* It is measured north (plus) and south (minus) from the equator. Declination on the sky is precisely analogous to latitude on the Earth.

The average plane of the Milky Way, or Galaxy, is indicated.

To Use the Star Chart

Face south with the chart before you.

If the time is 8 P.M., the constellations under the current month will appear in the South. The map will extend over your head and below the horizon.

At 10 P.M., look under the next month, at 12 P.M. under the second month, and so on. At 6 P.M., look under the preceding month.

If you are located in the Southern Hemisphere, hold the chart upside down and face North.

Appendix 5

The Planet Finder

Table 9 gives the positions of the brighter planets from 1981 through 1992. It was especially calculated for this volume from the tables by William D. Stahlman and Owen Gingerich (Solar and Planetary Longitudes for Years −2500 to +2000, Madison, Wis., 1963), courtesy The University of Wisconsin Press. It is designed for use with the accompanying star chart (Fig. 199). The numbers given are the celestial longitudes in degrees. The planets can be located near the ecliptic on the star chart at the corresponding values of the longitude. *Italics* are used for the longitudes of the *morning* sky.

The Sun's longitude is given for the evening of the 13th of each month. The superior planets, Mars, Jupiter, and Saturn, may be difficult to find when they are near to the Sun on the sky, that is, near conjunction; the date immediately preceding conjunction is indicated by a dagger †. They can be observed during the entire night when near opposition, indicated by an asterisk *. They are visible in the evening sky for about 2 months after opposition. Jupiter always lies within 2° of the ecliptic, Saturn 3°, and Mars 7°. For Mars this large deviation occurs only when the planet is near opposition; see Fig. 137 for the most favorable oppositions. Mercury moves so very rapidly that its longitudes are given at 10-day intervals. When no longitude is given in Table 9, Mercury is hopelessly close to the Sun. Mercury's dis-

TABLE 9. *The planet finder.*[a]

Planet	Sun	Mercury			Venus		Mars	Jupiter	Saturn
Date	13th	3rd	13th	23rd	13th	28th	13th	13th	13th
1981									
January	293	—	—	—	*272*	*292*	*313*	*190*	*190*
February	325	333	—	—	*312*	*330*	*335*	*190*	*190*
March	353	321	326	336	*346*	*005*	*358*	*187**	*188**
April	23	352	—	—	25	44	*21†*	*183*	186
May	52	—	—	84	62	79	*44*	181	184
June	81	93	—	—	101	119	*65*	181	185
July	111	—	90	103	138	155	*87*	183	186
August	140	—	—	—	174	192	*107*	188	188
September	170	180	195	207	211	228	*128*	194	191
October	200	213	—	—	245	262	*146*	200†	*194†*
November	231	*202*	*213*	—	277	292	*163*	207	197
December	261	—	—	—	303	309	*178*	212	200
1982									
January	293	—	—	—	306	298	*191*	*217*	*202*
February	324	—	*304*	*309*	*294*	*298*	*199*	*219*	*203*
March	352	*316*	*329*	—	*308*	*320*	*197*	220	202
April	23	—	—	—	*337*	*353*	*185**	*218**	199*
May	52	63	72	—	*9*	*27*	180	213	197
June	82	—	—	70	*45*	*63*	187	211	197
July	111	81	—	—	*81*	99	199	211	197
August	140	—	158	172	*118*	*137*	216	213	199
September	170	—	195	—	*157*	*175*	236	218	202
October	200	—	—	—	*194*	*213*	256	223	205†
November	231	—	—	—	233	251	280	231†	*209*
December	261	—	—	289	270	289	303	*237*	*212*
1983									
January	293	301	—	—	309	329	328	*243*	*214*
February	324	*290*	*299*	*312*	349	007	351	*248*	*215*
March	352	*324*	—	—	23	44	13	*251*	*231**
April	23	—	—	52	61	73	36	*251*	210
May	52	—	—	—	95	111	58	*248**	210
June	82	*49*	*59*	*73*	128	141	*79†*	244	209
July	110	—	—	—	152	159	*100*	242	209
August	140	153	167	177	157	*149*	120	242	211
September	170	180	—	—	*142*	*146*	140	244	212
October	199	—	—	—	*155*	*170*	159	248	215
November	230	—	—	—	*184*	*201*	177	254	*219†*
December	261	269	281	—	*217*	*236*	194	261†	222

a. Italics indicate that the planet is visible in the morning. All others are visible in the evening. † indicates conjunction with the Sun. * indicates opposition to the Sun.

Table 9. (*continued*)

Planet	Sun	Mercury			Venus		Mars	Jupiter	Saturn
Date	13th	3rd	13th	23rd	13th	28th	13th	13th	13th
1984									
January	292	—	_272_	_278_	_255_	_273_	_211_	_269_	_225_
February	324	_292_	—	—	_290_	_311_	_226_	_275_	_226_
March	353	—	—	—	_328_	_347_	_236_	_280_	_227_
April	24	_32_	—	—	_7_	_25_	_239_	_282_	_226*_
May	53	—	28	36	_44_	_62_	_231*_	_283_	223
June	83	_53_	—	—	_82_	100	223	280	221
July	111	—	132	146	119	137	224	276*	220
August	141	157	162	—	157	175	238	273	222
September	171	—	_153_	—	195	213	256	273	224
October	200	—	—	—	232	250	276	276	226
November	231	—	250	263	270	288	298	281	230†
December	262	276	—	—	305	322	321	287	_233_
1985									
January	293	_260_	_272_	—	340	356	344	_294†_	_236_
February	325	—	—	—	10	19	8	_302_	_238_
March	353	—	—	—	23	18	29	_308_	_239_
April	23	—	—	8	_9_	_6_	51	_313_	_238_
May	52	_16_	_29_	—	_11_	_21_	72	_317_	235*
June	82	—	—	—	_36_	_51_	92	_317_	233
July	111	125	137	145	_67_	_84_	_112†_	_315_	232
August	140	—	—	—	_102_	_120_	_133_	311*	233
September	170	—	—	—	_138_	_157_	_152_	308	234
October	200	—	—	229	_176_	_194_	_171_	307	237
November	231	243	253	—	_215_	_233_	_190_	310	239†
December	261	—	_241_	_251_	_253_	_271_	_209_	315	_243_
1986									
January	293	—	—	—	_291_	_310_	_228_	321	_246_
February	324	—	—	—	330	349	_246_	_328†_	_248_
March	352	—	—	—	5	24	_262_	_336_	_250_
April	23	_350_	_356_	_8_	44	63	_282_	_342_	_250_
May	52	_23_	—	—	81	99	_291_	_348_	_249_
June	82	—	103	116	118	135	_295_	_352_	244*
July	111	124	—	—	152	169	288*	_353_	244
August	140	—	—	—	186	201	282	_351_	244
September	170	—	—	—	215	226	287	348*	245
October	200	210	223	233	230	228	302	344	247
November	231	—	—	—	220	215	322	343	250
December	261	_231_	—	—	221	231	342	345	_253†_

a. Italics indicate that the planet is visible in the morning. All others are visible in the evening. † indicates conjunction with the Sun. * indicates opposition to the Sun.

Table 9. (*continued*)

Planet	Sun	Mercury			Venus		Mars	Jupiter	Saturn
Date	13th	3rd	13th	23rd	13th	28th	13th	13th	13th
1987									
January	293	—	—	—	*246*	*261*	3	350	*257*
February	324	—	342	—	*279*	*296*	25	356	*259*
March	352	—	*330*	*335*	*312*	*330*	47	2†	*261*
April	23	*346*	*0*	—	*349*	7	65	*11*	*262*
May	52	—	—	—	24	43	85	*17*	*261*
June	82	95	105	—	62	80	105	*23*	258*
July	110	—	—	*100*	99	118	124	*28*	255
August	140	—	—	—	*137*	156	144†	*30*	255
September	170	—	189	204	176	196	*164*	29	255
October	199	216	222	—	213	231	*183*	26*	258
November	230	—	*212*	—	252	271	*203*	21	260
December	261	—	—	—	292	307	*222*	20	263†
1988									
January	292	—	—	320	327	346	*243*	20	267
February	324	—	—	*314*	5	22	*264*	25	270
March	353	*317*	*327*	*340*	38	54	*283*	31	271
April	24	—	—	—	70	82	*304*	38	273
May	53	—	74	83	88	90	*324*	46†	272
June	82	—	—	—	80	73	*344*	53	270*
July	111	*81*	92	—	74	82	*1*	59	268
August	141	—	—	169	95	109	*11*	64	266
September	171	185	197	205	*126*	*142*	11*	66	266
October	200	—	—	—	*160*	*178*	0	66	268
November	231	—	—	—	*197*	*216*	0	62*	270
December	262	—	—	—	*234*	*253*	10	57	273
1989									
January	293	—	—	—	*273*	*292*	27	56	*277*
February	325	*297*	*299*	*308*	*312*	*331*	45	57	*280*
March	353	*319*	*334*	—	347	6	61	60	*282*
April	23	—	—	52	25	44	80	66	*284*
May	52	64	—	—	63	81	99	72	*284*
June	82	—	*60*	70	101	119	119	79†	*282*
July	111	—	—	—	138	156	137	*86*	279*
August	139	147	162	177	174	192	156	*92*	278
September	171	187	190	—	211	229	176	*98*	277
October	200	—	*182*	—	245	262	*196†*	*100*	278
November	231	—	—	—	278	291	*216*	*101*	280
December	261	—	279	291	302	308	*237*	*98**	283

a. Italics indicate that the planet is visible in the morning. All others are visible in the evening. † indicates conjunction with the Sun. * indicates opposition to the Sun.

TABLE 9. (*continued*)

Planet	Sun	Mercury			Venus		Mars	Jupiter	Saturn
Date	13th	3rd	13th	23rd	13th	28th	13th	13th	13th
1990									
January	293	—	—	*282*	304	295	*258*	92	*287†*
February	324	*290*	*302*	*314*	292	293	*281*	90	*291*
March	353	—	—	—	*307*	*321*	*301*	91	*293*
April	23	—	42	—	*337*	*353*	*324*	94	*294*
May	52	—	—	*40*	*10*	*27*	*347*	99	*295*
June	82	49	62	—	*46*	*63*	*9*	105	*294*
July	110	—	—	140	*81*	*99*	*30*	*112†*	*292**
August	140	156	167	173	*118*	*137*	*50*	*119*	290
September	170	—	—	—	*157*	*176*	*66*	*126*	289
October	200	—	—	—	*195*	*214*	*75*	*130*	289
November	231	—	—	—	234	252	*70**	*133*	290
December	261	271	280	—	271	290	59	*133*	293
1991									
January	293	*264*	*269*	*281*	310	329	57	*131**	*297†*
February	324	*296*	—	—	349	8	67	126	*301*
March	352	—	—	20	24	42	80	124	*304*
April	23	—	—	—	61	78	96	124	*306*
May	52	*18*	*25*	*38*	96	112	112	126	*306*
June	82	—	—	—	128	141	131	131	*306*
July	110	119	134	147	151	157	148	136	*305**
August	140	154	—	—	154	*146*	167	*144†*	302
September	170	—	—	—	*140*	*145*	188	*151*	300
October	200	—	—	—	*155*	*168*	207	*156*	300
November	231	239	252	262	*184*	*201*	*229†*	*161*	301
December	261	—	—	*251*	*218*	*236*	*251*	*164*	303
1992									
January	292	*261*	*274*	—	*255*	*274*	*273*	*164*	306
February	324	—	—	—	*293*	*311*	*296*	*161*	*311†*
March	353	—	—	—	*329*	*348*	*318*	*158**	*314*
April	24	—	*0*	*6*	*7*	*26*	*343*	155	*317*
May	53	*18*	*33*	—	*46*	*63*	*6*	155	*318*
June	82	—	—	114	82	100	29	158	*318*
July	111	127	135	—	119	137	*50*	162	*317*
August	141	—	—	132	158	177	*71*	168	314*
September	171	—	—	—	196	215	*90*	*174†*	312
October	200	—	219	233	233	251	*106*	*181*	311
November	231	245	249	—	270	288	*116*	*186*	312
December	262	*232*	*241*	*254*	305	322	*116*	*191*	314

a. Italics indicate that the planet is visible in the morning. All others are visible in the evening. † indicates conjunction with the Sun. * indicates opposition to the Sun.

tance from the Sun at greatest elongation varies enormously because of the eccentricity of its orbit. Each greatest elongation is indicated by at least one value of the longitude. When values of the longitude are given for four consecutive 10-day intervals in Table 9, there is a fair chance, with the naked eye, to identify Mercury on the sky. Choose a night near the middle of the series. When only one or two consecutive values are given there is little chance of finding Mercury without binoculars. When looking for Mercury it is best to estimate the longitude at the date between the values given. For the longitudes given, Mercury will always lie within 6° of the ecliptic.

The longitudes of Venus are given at 15-day intervals. The planet is so bright that it may occasionally be seen very close to the Sun at dates for which the longitude was not calculated, but at such times it will be visible only for a short time in the evening or morning twilight. Venus may lie as much as 7° from the ecliptic.

When their longitudes are not given in italics, Venus and Mercury can be seen in the evening. They can be observed in the morning sky only when the longitudes are in italics.

If instead of the star chart in Fig. 199 you wish to use some other chart, divide the numbers in Table 9 by 15, to convert them into hours of time. Then use the Right Ascension scale of the other chart and locate the planets near the ecliptic as with the present chart.

How to Find the Planets

Locate the year and the month in Table 9. Read the number in the table for each planet at the nearest date. Locate this number along the curved ecliptic on the Star Chart. Look for the planet near the ecliptic at this point. It is well also to locate the position of the Sun.

If no number is given in Table 9, the planet is too close to the Sun for observation. Do not look for a planet when the number carries a dagger (†) or immediately follows such a number. Note the italics *(000)* for planets in the morning sky. For about two months before an asterisk (*) in the table, a planet can be seen in the east in the evening.

Venus and Jupiter are brighter than any star. Mars may be a bit fainter than Sirius, the brightest star. Saturn is somewhat fainter but still bright. Mercury is very difficult to find.

Read Appendix 2 for definitions of planetary configurations and Appendix 4 for a description of the star chart.

Example: To locate the planets on November 5, 1986. In Table 9 for November 1986, we find that the entries are 231° for the Sun on November 13, none for Mercury, *220°* for Venus, 322° for Mars, 343° for Jupiter, and 250° for Saturn. Hence Mercury and Venus are too close to the Sun for observation, Venus being in the morning sky (italics); Mars and Jupiter are excellently placed in the evening sky, 91° and 112° east of the Sun, respectively. Saturn is rather close to the Sun (19°) in the evening sky.

Appendix 6

The Moon's Age

The Moon's age is measured in days from new moon; it is 7 days at first quarter, 15 days at full moon, and 22 days at third quarter. Table 10 is based on a simple approximate formula given by P. Harvey in the *Journal of the British Astronomical Association* in July 1941, and lists the Moon's age at the zeroth day of each month from January 1980 to December 2000.

How to Find the Moon's Age

Method A. From Table 10 (good from 1981 to 2000).

Enter Table 10 with the year and month. To the table entry add the number of the day of the month to give the Moon's age on that date. (If the sum exceeds 29, subtract 30).

The error is usually about 1 day, occasionally 2 days.

Example: What is the Moon's age on November 5, 1986?

For November 1986, Table 10 gives the entry 28. Thus the Moon's age on November 5, is 5 + 28 = 33 − 30 = 3 days. The Moon will be 3 days old and visible in the evening sky.

Note: For observations to be made in the evening, especially in the Americas, add 1 day to the age of the Moon determined from the table, since the dates here are for midnight at Greenwich, England.

Method B. For dates not included in Table 10.

Harvey's formula is usually accurate to within a day (or occasionally two) over the Christian Era, using the present Gregorian calendar. The calculation is as follows for November 5, 1986.

	For Nov. 5, 1986
Divide the year number by 19; keep only the remainder	10
	____
Multiply the remainder by 11	110
Add $\frac{1}{3}$ of the century number excluding fractions	+ 6
Add $\frac{1}{4}$ of the century number excluding fractions	+ 4
Add the number 8	+ 8
	____
Sum	128
Subtract the number of the century	− 19
	____
Difference	109
Add the number of the month, begining with March = 1 (February = 12 and January = 11 of the previous year)	+ 9
Add the day of the month	+ 5
	____
Sum	123
Subtract multiples of 30	− 120
	____
Age of the Moon (days)	3

TABLE 10. *The Moon's age on the zeroth day of each month (January 1981 to December 2000).*

	Jan	Feb	Mar	Apr	May	Jun	Jul	Aug	Sep	Oct	Nov	Dec
1981	24	25	25	26	27	28	29	0	1	2	3	4
1982	5	6	6	7	8	9	10	11	12	13	14	15
1983	16	17	17	18	19	20	21	22	23	24	25	26
1984	27	28	28	29	0	1	2	3	4	5	6	7
1985	8	9	9	10	11	12	13	14	15	16	17	18
1986	19	20	20	21	22	23	24	25	26	27	28	29
1987	0	1	1	2	3	4	5	6	7	8	9	10
1988	11	12	12	13	14	15	16	17	18	19	20	21
1989	22	23	23	24	25	26	27	28	29	0	1	2
1990	3	4	4	5	6	7	8	9	10	11	12	13
1991	14	15	15	16	17	18	19	20	21	22	23	24
1992	25	26	26	27	28	29	0	1	2	3	4	5
1993	6	7	7	8	9	10	11	12	13	14	15	16
1994	17	18	18	19	20	21	22	23	24	25	26	27
1995	28	29	0	1	2	3	4	5	6	7	8	9
1996	10	11	11	12	13	14	15	16	17	18	19	20
1997	21	22	22	23	24	25	26	27	28	29	0	1
1998	2	3	3	4	5	6	7	8	9	10	11	12
1999	13	14	14	15	16	17	18	19	20	21	22	23
2000	24	25	25	26	27	28	29	0	1	2	3	4

Index